First published in Great Britain in 1985 by:

Rosendale Press Ltd., Premier House
10 Greycoat Place, London SW1P 1SB

Revised edition 1994

Cover and book design by Pep Reiff
Printed and bound by G. Canale & Co. SpA, Italy

ISBN 1 872803 19 9

British Library Cataloguing in Publication Data.
A catalogue record for this book is available from the British Library.

CONTENTS

PREFACE

The first edition of this book and our interest in water stem from an article written for *Smithsonian* magazine in 1984. That initial assignment started us on a journey of discovery of the world's mineral and spring waters.

This revised and expanded edition covers the immense growth and changes in the bottled water business over the last decade. At the outset we have a special vote of thanks to Janet Long, who has worked with us on the new book, patiently revising original entries and tracking down new waters everywhere from Australia and Argentina to Finland, Poland, Romania and Turkey. In this pursuit, as with the first edition, we have often been guided by Dr Wilhelm Schneider of the Fresenius Institute in Germany. Richard Hall of Zenith International in Bath has also been most generous with his time and knowledge in leading us through a more complex international scene.

We must also acknowledge the help of Corinne Caspar-Bordaz, the International Marketing Coordinator at Nestlé Sources International in Paris, who guided us through their world-wide network, and Jane Lazgin, Director of Corporate Communications at The Perrier Group of America (also part of Nestlé). Henri Giscard d'Estaing, the Director-General of Évian, Badoit and the other mineral waters of Danone (formerly BSN), and Perrine Boivin brought us up-to-date on that group, while Richard Foulsham, former Development Director of Volvic in London, gave guidance on the French and British markets. Our thanks also to Patricia Carroll at the Natural Mineral Water Association in London, and to Dermot Byrne of the Bottled Waters Association in Ireland, who provided much useful information.

Lisa Prats at the International Bottled Water Association in Alexandria, Virginia, was, as ever, helpful on the American scene. And Becky Kent at Mountain Valley Spring Company in Hot Springs, Arkansas, gave us the latest on President Bill Clinton's favourite water.

In Sydney, Ian Brown, the Corporate Affairs Manager at Coca-Cola Amatil introduced us to the Australian water scene, covered for the first time in this edition. While from Oman, Saibal Sen, General Manager of the National Mineral Water Company, briefed us not just on that Sultanate, but the entire market in the Gulf. Back in Rome, Diane Seed, the food and drinks writer, commented for us on the new trends in Italy. In Frankfurt, Helga Filip again assisted us on the uniquely regional German market. From Buenos Aires, Mariano Castro, Technical Director of Aguas Minerales responded immediately with advice on Argentina, while Bija Cieszkowski helped to unravel the mysteries of some of the Polish waters. We are indebted to Sankichi Yamasaki at Yamasaki International in Tokyo for information on Japan, and to Adrian Feru in Bucharest for introducing us to the Romanian mineral water market.

Finally, we have enjoyed working with Pep Reiff, designer of the original edition, who gave this one its new look. Georgie Robins, Pamela Burden and Nasim Mawji in our own office have kept track of endless outgoing queries and inflowing bottles, labels and information.

MPG
TSG
London,
July 1994

Art Nouveau window at San Pellegrino depicts marble fountain in stained glass.

INTRODUCTION
WATER: THE BOOM

Drinking water is fashionable again. A century ago it was the custom to visit elegant spas to 'take the waters' and cure all manner of ailments. Today, in an era increasingly concerned about health, diet and potential pollution of municipal tap water, mineral waters and natural spring waters are back in vogue. The shift in drinking habits away from strong spirits to lighter wines and from sugary soft drinks to bottled waters is world-wide. 'Water (consumption) is growing and outpacing all other beverages on a global scale,' says industry analyst Richard Hall of Zenith International. 'In Europe it's already a bigger market than carbonated soft drinks.' Around the world, over 42 billion litres (11 billion US gallons) of bottled water was drunk in 1993, worth $14 billion wholesale.

Such pickings have made water big business. A decade or so ago, the bottling of mineral and spring waters was largely a regional, often a very local affair. Although the French were already getting into their stride carving out a global image for mineral waters like Perrier and Évian, and owned a few other springs here and there, there was no true multinational. Today Nestlé Sources International, which has taken the Perrier group under its wing since 1992, controls 55 bottled water brands in 16 countries. They reckon to control around 15 per cent of world bottled water sales, which brought them revenues of close to $2 billion in 1993. Hard on Nestlé's heels is its great French rival Danone, which not only owns such flagship French waters as Évian, Badoit and Volvic, but has well-known waters like Ferrarelle in Italy and Font Vella in Spain. And, most recently, the Germans have joined in the international pursuit, with Gerolsteiner (already the leading brand there) snapping up an Italian group, Terme di San Andrea, with nine sources to its credit.

The attraction is not just the rate of growth – consumption has more than doubled in virtually all industrial countries in the last decade – but the prospects in emerging markets of Eastern Europe and Asia, to say nothing of China and the Commonwealth of Independent States. Before the Berlin Wall came down, consumption in what was then East Germany was a mere 5 litres a head; today it is 35 litres. Poland, Romania and other eastern European countries (whose waters are included in this guide for the first time) are reviving bottling at famous sources. Conglomerates like Nestlé are circling. 'We're looking at all of eastern Europe and Russia,' Nestlé's chief executive officer, Serge Milhaud, told us in Paris. 'The markets there will progressively adapt to a new way of life.' Already Nestlé is developing Naleczowianka in Poland (and in Asia has launched La Vie in Vietnam). The potential is clear. A pie chart of world-wide consumption shows Europe with a huge slice at 64 per cent; the United States with 21 per cent, the Far East with 3.5 per cent and 'others' (eastern Europe, Latin America, the Middle East and Africa) with around 11.5 per cent. The pie will be much more evenly sliced by the year 2000.

To many Europeans, of course, this is nothing new. 'Mineral water is deep in our culture,' said Évian's Henri Giscard d'Estaing. A bottle of wine and of mineral water complement a good meal in a restaurant. The connoisseur will choose the right water for the right occasion. Thus the sharp sparkle of Perrier or Apollinaris makes a good aperitif; the lighter natural carbonation of France's Badoit,

"And which particular water do you identify yourself with?"

Italy's Ferrarelle or Belgium's Bru goes well with food; the pure, still waters of Évian, Vittel or Volvic in France, Panna in Italy or Font Vella in Spain are not only good table waters, but suitable for mixing baby formulas. Chic French women, anxious to preserve their figures, sip the more highly mineralized Contrex as a diuretic.

These are all mineral waters of the classic European tradition. They are bottled at source without treatment under tight regulations, now harmonized in the European Union by a rule book known as directive 777 (see pages 194/195). Elsewhere, clear regulations often do not exist. In the United States, only the state of California has a precise code, although new federal regulations are due in 1994.

Thus outside Europe, the language of the label can be a minefield. Is it really a mineral or spring water? Or is the word 'spring' artfully in the name? Debates, even legal battles, rage in the United States on whether 'spring' means only free flowing waters bubbling from the ground, or covers, too, a borehole drilled to an aquifer (which is widely accepted).

The majority of waters in this guide are *bona fide* mineral or spring waters, bottled at source, but we have broadened the definition to meet local customs and needs.

What the papers and people say...

"The people who say these things about me are the ones who drink my Évian water with me and call themselves my friends. But that's Hollywood." Steven Spielberg

"Water snobbery has replaced wine snobbery." Time

"Water is now big, big, big business." The Observer

"More people are apparently craving something simpler; plain, unflavoured, non-carbonated water." Wall Street Journal

"Mineral water has been accepted as an alternative to having a drink and still being part of the party" John Hamilton, Gleneagles Water

"Water ... is so pure and clean and so Nineties." The Independent

WATER: THE ORIGINS

Natural mineral and spring waters owe their properties primarily to the rock strata through which they percolate. Water that first falls as rain or snow is both filtered by and picks up minerals from the various rocks through which it passes. Limestone will make a water rich in calcium, dolomites will contribute magnesium, and igneous rocks resulting from volcanic action will impart sodium. The process may take years so that the water matures, in a sense, like wine. Évian, for instance, estimates that it takes well over 15 years for water from the Alps to reach its source.

Although many minor elements are often present, the dissolved minerals in natural waters comprise largely calcium, magnesium, sodium, potassium, bicarbonate, carbonate, sulphate, chloride and nitrate. Iron may be present in suspension, but most is normally filtered out before bottling. Among the minor or 'trace' elements, one of the most common is fluoride, which, in moderation, is good for teeth. The basket of minerals in each water is judged by the total dissolved solids (tds) or dry residue obtained by boiling the water and heating the residue at 180°C (356°F).

The precise cocktail varies dramatically depending on the geology of the region. We encountered an Austrian water of volcanic origin, Sicheldorfer, so laden with sodium and bicarbonate that it contains 5,400 milligrams per litre of mineral salts, while in France we drank Charrier which has a mere 12 milligrams per litre and may almost be regarded as a distilled water. Broadly speaking we have defined a lightly mineralized water as one with less than 500 milligrams per litre, medium mineralization as up to 1,000 milligrams per litre and high mineralization as over 1,000 milligrams per litre. But precise definitions vary from country to country.

High mineralization is most frequently encountered in warm, very deep waters, usually occurring in regions of volcanic activity. They may also emerge naturally effervescent. Some of these thermal waters have such a high mineral content that they may be drunk only under medical supervision since they may cause quite violent intestinal reactions; a few are so potent as to be suitable only for bathing. Those bottled commercially are always the milder varieties, like Vichy Celestins from France or Ferrarelle from Italy, which may be drunk quite safely without limit.

By contrast, many mineral waters emerge as pure, cool, natural springs. Rain or melted snow simply filter through permeable soil or rock until they are stopped underground by impervious rock, such as granite or clay. Shallow springs emerge at the junction of permeable and impermeable rocks close to the surface. But mineral waters usually come from much deeper underground artesian aquifers from which the water rises through fissures under the pressure created by superimposed impermeable layers. And the very fact that the aquifer is thus sealed off above from ground water (which may well be polluted) maintains the purity and unchanging mineral content.

Whether they are thermal or cool springs, the sources are usually capped – to prevent any outside pollution getting in. Frequently boreholes are drilled down adjacent to the original spring (or galleries are built into a mountainside) to tap the aquifer directly. But the natural flow must not be exceeded in bottling for fear of disturbing the whole water table. The credentials of a mineral water are its constant temperature and chemical composition; to draw it off too rapidly could upset the delicate balance of minerals assimilated over years underground. In France, for instance, the Ministry of Mines designates the maximum amount of water that may be taken off each year to preserve the water's identity. The next criterion of a true mineral water is that it is bottled at the source, without any treatment, except filtration to remove iron and sulphur (which would discolour bottles). It should not be pasteurized, ozonated or subjected to ultraviolet light to render it sterile, but bottled under conditions of the most stringent hygiene. Indeed the whole essence of a mineral water is that it is 'live'; that is to say, it contains bacteria. This is no cause for alarm; rather that is where the real benefit comes from. It's just like eating yoghurt, which contains benign bacteria (patients who have undergone a course of antibiotics are often advised to eat yoghurt to restore vital bacteria to the system). The therapeutic claims for mineral waters rest partly on the fact that they are 'live'; for this reason they should be drunk within a day or two of breaking the seal, because once a bottle is opened other bacteria will enter.

Grand staircase in the casino at San Pellegrino.

WATER: THE TRADITION

The traditional reason for drinking mineral waters, especially in Europe, was precisely for their mineral content as an aid to health. Waters gained their reputation for specific benefits. Some highly mineralized waters, like Contrexéville in France, Fiuggi in Italy and Radenska in Slovenia, were famous for helping to break up kidney stones and curing urinary complaints. Others predominantly known for their bicarbonate level, like Fachingen in Germany, Ferrarelle in Italy and Vichy Catalán in Spain, eased digestion. Still others, like Apollinaris, helped bronchial complaints or, like Évian, soothed skin diseases. The diversity of waters meant that people took the one whose blend of minerals, or main mineral characteristic, best suited their complaint

The beneficial properties of many European waters enjoy a long history. Successive Roman emperors had a rare predilection for mineral waters both at home in Italy and on their conquests abroad. They established spas in France at Badoit and Vittel, while Julius Caesar took the warm waters at Vichy, which was known as Vicus Calidus (the hot town). Pliny the Elder in his *Natural History* reported not only on the 'miraculous waters' of Ferrarelle, but added that those of Spa, near Liège, Belgium, were 'already famous'.

Testimonials to delight the modern press agent kept arriving over the centuries. Leonardo da Vinci regained his health with the waters of San Pellegrino in northern Italy and Michelangelo went in to bat for Fiuggi. Tormented by kidney stones he sampled that water and soon wrote, 'I am much better than I have been. Morning and evening I have been drinking the water from a spring about 40 miles from Rome, which breaks up the stone … I have had to lay in a supply at home and cannot drink or cook with anything else.'

Russia's Peter the Great found that the effervescent waters of Bru in Belgium so eased his indigestion that he gulped 21 glasses each morning. and the German writer, Goethe, settling in to sample his local waters at Fachingen, wrote to his daughter-in-law, 'the next four weeks are supposed to work wonders. For this purpose I hope to be favoured with Fachingen water and white wine, the one to liberate the genius, the other to inspire it.'

The bottling and export of favoured waters in earthenware jars carefully packed in straw began remarkably early. Those from Spa in Belgium were going to the capitals of Europe in the 16th century, while an Italian water, known as Acqua dei Navigatori, was used to stock boats setting off for the New World. And by the 19th century many French, Italian and German waters were widely distributed. Louis Pasteur, the French scientist, regularly sent to Badoit for 10 cases at a time, and a glass of Vichy water sipped in the French colonies became a refreshing reminder of home. Apollinaris from Germany became fashionable in England, and was a regular drink on many railroad cars out west in America.

The railway age, which made it easier to distribute waters, also enabled the wealthy to travel long distances to visit spas. Grand hotels and ornate drinking halls in which to sip graciously and promenade, together with casinos and theatres to while away the evenings, were built. Spas such as Contrexéville and Évian in France or Boario Terme and San Pellegrino in Italy became part of the social scene. In the United States the fashionable venues included Hot Springs, Virginia, and Saratoga Springs in New York state.

French executives relaxing at Contrexéville.

The faded grandeur of that era is still to be found at many resorts. Although some now boast only boarded up hotels, at others the season is still busy. The French, in particular, have made a spirited attempt to polish a modern image for their spas, appealing not just to jaded executives, but to health-conscious families. Contrexéville now calls itself 'the slimming capital of Europe' attracting smart Frenchwomen to drink its diuretic waters and jog in its *parc,* while the neighbouring Vittel has turned over its Grand Hôtel to the lively club Méditerranée.

At each spa, a modern combination of light, pleasant, three-course meals (this is France after all) and exercise in the open air has been linked to immersion in mineral water lore. On arrival, the slimmer is welcomed to the hotel and blood, urine and other tests are taken to establish basic metabolism. A doctor then prescribes the appropriate water. At Contrexéville, for instance, he may recommend either Source Pavilion, which is the bottled water, or Source Souverain, which is more laxative and not sold to the general public. Each morning the slimmer is woken with a glass of freshly drawn mineral water. Mornings are then spent in sessions of hydrotherapy, gymnastics or relaxation classes until 11.30, when everyone gathers at the source for another glass of water. The spas offer tennis, golf, swimming, cycling and jogging.

WATER: THE NEW IMAGE

On the catwalks of Paris fashion houses in the spring of 1994 there was a new look. 'The next hot fashion accessory is a bottle of water,' bubbled Marion Hume, fashion editor of *The Independent*. 'So puritanical is fashion now that bottles of Vittel, Évian and Glacier (the last is particularly stylish because the bottle is designed by Philippe Starck) have replaced the statement jewellery, handbags and scarves.' The bottle of water, Ms Hume observed, has also become *de rigeur* for chic window dressing. Ralph Lauren's Polo Sport shop in New York is decked out with Évian to present a new message of health and body consciousness.

Fashion is almost a late-comer. Watch Wimbledon, the World Cup Final, the Open Golf Championship or indeed any great sporting event, and everyone is imbibing by the litre. President Bill Clinton sustains himself with Mountain Valley Spring while out jogging; Princess Diana slips out of her health club with a bottle of Volvic. Debate at any summit meeting anywhere is cooled by all those bottles neatly assembled before presidents and premiers.

The Romans may have started it all 2,000 years ago, but the French have taken over. They are the driving force, the impetus behind promoting mineral and spring waters throughout the world. Who is sponsoring Puccini at the great Sydney opera house? Vittel. Who paid for 16-page advertising inserts in *Rolling Stone* magazine and *The New Yorker* in 1993? Évian. Even the Gulf War was not lost as an opportunity. US troops out in the desert were seen on the nightly news shows refreshing themselves with Évian. Perrier, of course, got it all going with those punning slogans of the early 1980s. 'Eau La! La!, 'H_2 Eau', 'Bistreau' and 'Anything else is pseudeau', which proved water could be advertised internationally, and did much to get the Americans and the British drinking it.

Is it audacity to persuade people to pay money for water? Not if the water is good enough, seems to be the answer. Perrier became the alternative to three martinis before a New York lunch, and no one minded paying $5 in a bar for the club-shaped bottle. Even in the City of London, where merchant banks often have wonderful wine cellars for their private dining rooms, bottled water is now passed round more than the claret or port. Even in the dealing rooms, fevered traders have a bottle to hand.

Mineral water is the health drink discovered by the age that has turned to health foods. It has acquired a sporting image - almost like internal aerobics. We are concerned to cut down not just on alcohol but on caffeine, sugar and sodium. Water fits the bill. In the United States, new Food and Drug Administration rules require food and beverages to declare 'nutrition facts'. Typical water labels happily proclaim: 'Calories 0; Total Fat 0; Sodium 0; Total Carbonates 0; Protein 0' – a perfect scorecard. Women are constantly advised to drink more to avoid problems like cystitis, and here is a calorie-free way. A reasonable level of fluoride in daily water intake protects the teeth; many mineral waters contain a useful dose. Calcium for strong bones is needed not only by children, but also by women to avoid post-menopausal osteoporosis; this is available in Perrier, Vittel, San Pellegrino, and almost any German or Swiss mineral water.

While many of us may owe our lives to antibiotics, we may still be wary of turning to these powerful weapons too often. Mineral waters, especially in France, Italy and Germany, remain as a natural alternative medicine. When we visited Vichy in France, our young

guide was convinced that his recent recovery from a kidney infection had been achieved with less damage because he had taken a cure with the local waters and not used modern drugs. A glass of mineral water containing natural bicarbonates with a meal also makes more sense than downing antacid tablets later.

Moreover, mineral water tastes so much pleasanter than the water from the municipal tap. Travelling for this new guide, we have found concern about tap water stimulating the swing to bottled everywhere. Tap water may be 'safe', but what are the dangers of drinking heavily chlorinated water for a lifetime? The water authorities face a dilemma. Tap water has to fulfil the disparate needs of drinking, cooking, bathing, doing the laundry and washing the car. Yet a hard water is better for drinking, while a soft water is essential for a good lather. In many places a mere 2 per cent of the water is drunk; 98 per cent fulfils other needs. Increasingly, the consumer's answer is to wash in tap water, and drink bottled. Which is precisely why it has become big business.

Auguste Saturnin Badoit, first bottler of Saint Galmier water.

FRANCE

INTRODUCTION

The attention that the French lavish upon their waters matches the loving care with which they make wine. And a bottle of water and a bottle of wine are natural accompaniments to a good French meal. France, consequently, is the unquestioned leader in bottled mineral waters. It is French mineral waters such as Évian, Perrier and Vittel that are to be found in restaurants and hotel mini-bars on every continent. Often their parent companies, the water giants Nestlé Sources International and Danone (formerly BSN) both based in Paris, own many local waters too. Nestlé alone now owns 55 major springs in such diverse countries as Belgium and Brazil, Germany and Greece, Italy, Poland, the United States and Vietnam; it reckons to control 15 per cent of the entire bottled water market worldwide.

Traditionally, the French looked to mineral waters for therapeutic treatment, and the reputation of several of today's most famous sources was already established when the Romans ruled Gaul. Remains of Roman spas turn up at Badoit, Perrier and Vittel. Julius Caesar took the waters at Vichy, which was known as Vicus Calidus (the hot town). Regulation also began early. The first decree governing mineral waters was issued in 1781, and in 1856 came the crucial law enabling the state, after careful testing, to declare a source *d'interêt public*, indicating that its waters were beneficial to health. That essential credential, *source declarée d'interêt public*, initially encouraged people suffering from all manner of maladies to take the cure at Évian-les-Bains or Contrexéville or Vittel. Vichy became a resort for French colonials, on home leave from the unhealthy climates of Africa and south east Asia, to restore their health drinking the waters. And the waters that were bottled and distributed were primarily to enable people to continue their treatment when they returned home, or were bought by others to help some ailment. Before the Second World War, if a French child had digestive problems, he was given a dose of Vichy water.

For all that loyalty, the French now rank only fourth in the European league for drinking mineral waters; they have been overtaken by the Italians, Belgians and Germans. The French each consume just over 80 litres (21 US gallons) of mineral waters annually, although if *les eaux de source* (natural spring waters) are included the total virtually matches the Belgians at 103 litres overall. The strong preference, incidentally, is for still water, which accounts for 80 per cent of the market. What makes France unique, however, is the devotion to a mere handful of nationally distributed mineral waters, still and sparkling. Contrex, Évian, Vittel and Volvic command virtually 60 per cent of the still market; while just Badoit and Perrier enjoy almost 60 per cent of sparkling sales, and that rises to 75 per cent if the powerful brew of St. Yorre is added in. But it is that national popularity, contrasting with the regional nature of bottled waters in most other European countries, that has given the French waters their springboard to make the leap abroad. Even before new alliances in the 1990s, the French companies had the resources to export. In Britain and the United States the campaigns of Perrier in the early 1980s gave the impetus for growth. So in the mid-1990s the French export over 1 billion litres (262 million US gallons) of their waters, more than any other country.

Expansion, however, has not lifted the strict obligations imposed by the Ministry of Health, the Academy of Medicine and the Ministry of Mines on mineral waters tagged *d'interêt public*. While the French have agreed to abide by the European Union directive 777 that guides them all, the French set their own eight basic rules long ago:

1. The mineral water must be recognized as therapeutic by the state.
2. The mineral water must come from one underground reservoir alone, and must emerge with constant composition and temperature.
3. The mineral water must not be treated in any way (although filtering to remove iron is

Sipping Vichy waters was always associated with fashion.

permitted in special circumstances).

4. The mineral water must be bottled at or immediately close to the source.
5. Not only must strict hygiene be observed, but the entire operation is subject to constant checks by the Ministry of Health. The Pasteur Institute makes analyses of the water every two months.
6. The Ministry of Mines lays down a maximum daily flow that may be tapped to ensure that the water table is not disturbed.
7. Other drinks (such as soft drinks) may not be made in the same bottling plant.
8. A protected perimeter, usually 6 or 8 kilometres (3 or 4 miles) in radius, is established around the catchment area and no underground work of any kind may be undertaken without a special study by the Ministry of Mines.

Also, a pre-condition for mineral waters to be bottled and sold commercially is that they can be drunk in unlimited amounts; there cannot be a daily limit as there would be if they were medicinal. This requirement means that, of 1,200 mineral waters known in France, scarcely 50 have been granted the accolade *d'interêt public*, and thus may be bottled. The remainder can be taken only at the spa itself under medical supervision.

The mineral water classification, however, is supplemented by a lesser *appelation*, that of *les eaux de source*, essentially spring waters which are also subject to government regulations. They, too, must have constant physical and chemical characteristics, be biologically pure, be bottled at source without any treatment and be clearly labelled with an *appelation d'origine*. That is to say, spring waters from several different regions may not all be bottled under a single brand label. A water must come from one area, just as a wine like Beaujolais must come from clearly specified vineyards. *Les eaux de source*, however, are not billed as having any beneficial therapeutic properties; they are simply refreshing spring waters. And their popularity is growing rapidly, since they are cheaper than the great names of Contrex, Évian or Vittel, because they are bottled and sold locally. Transport costs are a headache for the big mineral water groups, whose sources are primarily located in the three mountainous areas of the Alps, the Massif Central and the Vosges, far from the main cities. They often cannot compete in price on hypermarket shelves several hundred miles away with a local *eau de source*.

Over 50 *eaux de source* are recognised in France and their sales are approaching 1.5 billion litres (396 million US gallons), compared to over 4 billion (1,056 million US gallons) for domestic mineral water sales.

Despite this challenge, the taste for mineral waters in France is unabated, although there has been a shift in favour of lower mineralization. Moreover, it comes as something of a surprise to a foreign observer to find that the concept of drinking mineral waters, including those highly mineralized ones flowing only at the spas, still occupies an important medical niche.

For the French *la médecine douce* (alternative medicine) is widely accepted and is often paid for by social security. Doctors can prescribe 'taking the waters' to aid kidney and digestive ailments, rheumatism and arthritis; physiotherapy and massage with waters is also sometimes recommended after accidents. But apart from the true *curistes*, thousands of other French men and women, many of them quite young, pay their own way for a 10-day health package at one of the spas. At Vittel, Club Méditerranée, that epitome of easy-going vacations in the sun, is running the old Grand Hôtel as a health farm for jaded executives. But at the spas themselves, people drink from a selection of different waters, more varied in mineral content and temperature than the water selected for bottling. At Vichy, apart from the cool water of Source des Celestins, which is bottled, there are the hot waters of Source Hôpital, Source Chomel and Source Grande Grille, and the cooler waters of Source du Parc and Source Lucas; the precise quantities of each, to be drunk at carefully charted intervals during the day, are prescribed individually. The busiest time in the drinking halls of Contrexéville, Évian, Vittel or Vichy is between 11.30 am and noon, when a ceaseless *passaggiata*, equipped with glass drinking mugs, marked in centilitres and housed in little wicker baskets, lines up at the *gryphons* through which the water flows ceaselessly. Everyone carefully fills up to the recommended level, and then goes out to sit in the sun, sipping judiciously and chatting to friends. The image of spas as modern fitness centres has been encouraged by the bottlers. The aim is to project spas, not as rest homes for played-out colonials, but for sporty, health-conscious families. When we first went to Vichy, where Perrier had invested millions to make it a more swinging town, we were constantly urged to try the tennis courts, golf courses and night clubs.

Badoit benefits were translated, not quite accurately for English consumption. Poster reflects regal society that 'took the waters' at Evian in 1925.

In those days Vichy waters were under the umbrella of Société Anonyme Source Perrier, along with Perrier itself, Contrexéville, Volvic and an assortment of regional mineral waters. The creator of this empire was Gustave Leven, a former stockbroker, who plotted every move for nearly 40 years. Under Leven's energetic direction Perrier began the great world-wide promotion of its sparkling water. When we first visited the Perrier source in 1984, 45 per cent of its water was already being exported to 150 countries. Not content with exporting French waters, the group was also buying up springs everywhere abroad.

Nowadays it requires an odyssey to 55 waters in 16 countries to take in all the springs owned by Nestlé Sources International, the holding company created in December 1992, to embrace not only Perrier but Vittel too, along with assorted water groups such as Blaue Quellen in Germany, that Nestlé itself already controlled. It notches up new acquisitions almost monthly: Korpi in Greece, Minéré in Thailand, Deer Park in the United States, La Vie in Vietnam, Santa Maria in Mexico and Naleczowianka in Poland.

Hydrotherapy at Vittel.

This empire, however, is still controlled from 18 Rue de Courcelles, in Paris. The new chairman and chief executive is a quietly spoken, courteous man, Serge Milhaud. The aim, he explained, is simple. 'To proceed on two levels. First, to reinforce the positions of the international brands – Perrier, Contrex, Vittel – and secondly the acquisition of new springs, either undeveloped or already being developed.' He reminds visitors, with a wry smile, that 'a spring cannot be moved'. The major springs of world-wide fame are in France; there they stay. So to participate in competitive local markets, springs, too, must be acquired.

Actually, in the grand reshuffle of French waters in the early 1990s, Nestlé has come out with a rather different hand than Perrier originally held. That in itself has dictated one of Milhaud's first main moves on home ground. Nestlé has Contrex and Perrier, along with the several waters of Vittel which it also took over. But it had to give up Perrier's hold on the waters of Vichy (now owned by the Castel group) and, perhaps more significantly, Volvic. Although this left Nestlé with around 30 per cent of both still and sparkling water markets in France, the loss of Volvic deprived Nestlé of a lightly mineralized still water. 'In the last 15 years,' Serge Milhaud told us, 'the expansion in France has been in lightly mineralized waters and *eaux de source* – the rest of the market in still, highly mineralized waters, like Contrex, or sparkling, like Perrier, has been relatively stable.' So Nestlé needed a competitior to Évian, which enjoys the largest share of 'light' sales. Fortuitously, the Perrier group, prior to the take-over, had been scouting for new springs in the forests of the Ardennes, just across the Belgian border from north east France; a new water, with copious flow, was there, Valvert. Thus, in 1993, Nestlé took the plunge, launching Valvert as the first major new still mineral water (albeit a Belgian one) on the French market for 25 years. Milhaud admits it is a big challenge; Évian is a name known not just in France but around the world.

And what does Évian think? Évian is the flagship water of the French food and beverage group Danone (known as BSN until July 1994), whose other activities include everything from beer to yoghurt. As it happens their offices in Paris are just around the corner from Nestlé Sources International. The approach at Danone is somewhat different, according to Henri Giscard d'Estaing, director-general of their mineral water division. The philosophy is not the pursuit of ever more springs in ever more countries. 'We have exceptional waters, with exceptional names, with ample reserves,' said Giscard d'Estaing, 'so we consider we have plenty to do with them.' Indeed, he has a handful of the best 'château bottled' waters. Along with Évian, the Danone group owns Badoit, the lightly effervescent water that goes excellently with meals, which enjoys nearly 30 per cent of sparkling sales in France; and Volvic, which it acquired in the grand Perrier reshuffle. Abroad, too, it has famous names: Font Vella, the best-selling still water in Spain, and Ferrarelle, the top sparkling water in Italy. 'As with wine,' observed Giscard d'Estaing, 'there is room for the château-bottled, *grand crûs* in almost any international market.' The motto is quality rather than quantity. But Danone cannot rest entirely on its laurels. Badoit is such a popular water that over the last decade the source has reached the limit at which water can be taken off, without affecting the quality or the water table. With Badoit at the limit, Danone set off in search of new sparkling mineral waters; it has come up with a brace: Salvetat from a mountain park in Languedoc and Arvie from the Auvergne region. Salvetat, billed as *l'eau du soleil* (water of the sun) has more sparkle than Badoit, so that it can be drunk generally for refreshment, not just with food. Arvie, launched only in 1994, is more highly mineralized with a distinct flavour that suggests the strong waters of Vichy, without being quite so potent. 'We think there is room for a water with flavour,' explained Giscard d'Estaing.

Nestlé and Danone are in a league of their own. The only other significant group is Castel which, besides taking on the Vichy waters, St. Yorre and Vichy Celestins, had made a speciality of developing regional mineral waters, such as Charrier, Chateauneuf, Thonon and Vernière. The Vichy waters do give Castel almost 20 per cent of sparkling sales, but that is the smaller segment. The battleground in France is always in still waters. That is where regional mineral waters not only hold their ground, but seem to be gaining. They have the advantage of cheap local deliveries, but must bottle water to the same exacting standards. So, if you are in the south, look out for Pioule, in Brittany expect Plancoët, near Bordeaux you may encounter Abatilles. But when it comes to price, the big names do compete very hard. One of the most impressive sights at Évian is nine railway tracks at one end of the huge bottling plant, from whence hundreds of railroad trucks speed the bottles all over France.

The French continue to buy their water as readily as bread or wine. Some years ago Paul Bordier of the French Mineral Water Association suggested to us, 'Why not drink a litre of marvellous water a day? It costs half the price of your newspaper.' It still does: *Le Figaro* is 6 francs, a litre of Évian is 3 francs or less.

National Association:
Chambre Syndicale des Eaux Minérales,
10 rue de la Tremoille,
75008 Paris,
France.

Badoit: naturally lightly carbonated, medium mineralization.

BADOIT

This is the gourmet's mineral water, seen on the tables of most restaurants in France as an accompaniment to good food and wine.

As Badoit water pushes its way up from a deep water table through a 500-metre (1,635-foot) fissure in the local granite, it surfaces with a light natural sparkle and a measure of sodium bicarbonate and fluoride at a constant temperature of 16°C (60.8°F). The fluoride has ensured that the children of Saint-Galmier, where Badoit emerges, have half as many dental caries as children a mere 10 kilometres away. Badoit enthusiasts much further afield swear that the gentle sparkle and bicarbonate have a lightening effect on even the richest meal.

Workers in the 19th-century fill bottles at bottom of 27 metre (90 foot) well.

Known to the Romans (there are ruins of a Roman bath nearby), the water of Saint-Galmier in the dry hills near Lyons in central France, midway between the Massif Central and the Alps, was used and recommended by local doctors in the 17th century. The water was first ever bottled in France, when the spring was bought by Auguste Saturnin Badoit in 1837, on condition that the local population could always fill their bottles freely. He was soon dispatching his bottles, then filled by hand at the bottom of a 27-metre (90-foot) well, to Paris, Geneva, Marseilles and even Algiers. By 1859, the Badoit plant was producing 1 million bottles a year. On file is a letter from Louis Pasteur, the father of modern hygiene, writing from Italy in 1887 begging for the dispatch of 50 bottles of Badoit and complaining that customs officers had broken a couple of bottles in the last shipment. At the time of the great Paris exhibition at the turn of the century, Badoit was producing 15 million bottles a year, still packed with wooden stoppers, and was the only mineral water sold on such a scale in France at that time. The Badoit spring was declared *d'intérêt public* in 1897.

Today, Badoit water is tapped from a drill 153 metres (500 feet) below ground in sterile conditions, without any tampering with its carbon gas. The trick, we learned when we first went round, is to prevent any disturbance of the water to stop the carbon gas being pushed out. The water therefore travels slowly along stainless steel piping into the bottling plant. Badoit contains a mere 2.5 grams per litre of carbon gas, so lightly carbonated that it can be bottled in tough plastic. In 1973 it became the first carbonated water to be bottled in plastic and its distinctly designed bottle is made at the Saint-Galmier factory where one machine alone turns out 27,000 bottles an hour. The factory is pressurized to prevent any pollutants entering the atmosphere.

Badoit's very success, however, has created a problem. The source has reached the limit, about 260 million litres (69 million US gallons), that may be extracted in any one year from the narrow granite fissure. Consequently, exports to countries like Britain have been curtailed; the treasured water is kept at home for the French to enjoy with their meals. And Danone, the food and drinks group which owns Badoit, has launched two other sparkling waters, Salvetat and Arvie to maintain their share of the growing market.

Production: 260 million litres (69 million US gallons).

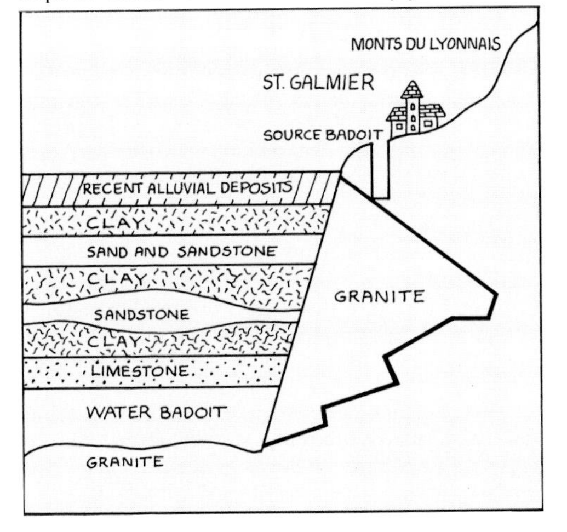

Bottling company:
Source Badoit,
SA Eaux Minérales d'Évian,
42330 Saint-Galmier,
France.
Owned: Danone (BSN).

Analysis:	*mg/l*
Calcium	*200.0*
Sodium	*160.0*
Magnesium	*100.0*
Potassium	*10.0*
Bicarbonates	*1,410.0*
Sulphates	*33.0*
Fluorides	*1.0*

Villagers fill up bottles at the public fountain outside Badoit's head office.

Contrex: still, high mineralization.

CONTREX

Contrex is known as the slimmer's mineral water, and sold with a heavy emphasis on its diuretic properties. One of France's top sellers, with around 15 per cent of the still market, it is nevertheless seen more in French homes as an all-day drink than on restaurant tables.

The town of Contrexéville itself in the Vosges hills, 350 kilometres (217 miles) east of Paris, is an established spa. The mineral springs at Contrexéville were discovered in the 18th century by the court physician of Stanislas, the Duke of Lorraine, also king of Poland and stepfather of Louis XV.

Two Romanoffs buried in the Russian chalet in the park surrounding the springs testify to the popularity of Contrexéville as a spa in its late 19th-century heyday, when it attracted nobility from all over Europe. The casino with its tiny theatre was, and is again today, a centre of entertainment. Contrex's springs were declared *d'intérêt public* in 1861.

In the park surrounding the drinking pavilion visitors enjoy the sun.

Dieters refill their mugs from the pavilion's silver tap at regular intervals.

Today only about 2,000 come to drink the Contrexéville mineral waters as a cure for kidney stones or gout, or other intestinal disorders. Many more, especially fashionable women, come to take part in a 10-day diet – exercise and two-and-a-half litre-a-day drinking programme as part of a slimming schedule. After the jogging and aerobics classes, slimmers can be seen regularly filling their glass mugs, marked in centilitres, at the exotic blue-tiled pavilion in the park.

The waters of Contrexéville, which take two years to filter down through layers of limestone into the immense underground reservoir from which they spring at a temperature of 11°C (51.8°F), are high in magnesium, calcium and sulphates. The water that comes out of the main silver *gryphon*'s head in the pavilion, Source Pavillon, is the same water which is bottled in the modern factory up on the hill.

For many years the labels read Contrexéville: Source Pavillon. Now the water is marketed simply as Contrex, subtitled with Source de Contrexéville on French labels and Source Pavillon on export labels.

The water, now owned by Nestlé, originally came under the wing of Perrier in 1954, achieving widespread popularity in France to reach a peak production of 850 million litres (225 million US gallons), virtually all of it in plastic bottles, usually in 1.5 litres.

Today's factory, where the plastic powder to manufacture the compound for the bottles arrives at one end and crates of bottled water are automatically shifted on to railway lines entering the factory at the other, presents a picture of highly controlled and integrated efficiency.

As the laboratory director remarked, 'Natural mineral water must be respected. You can't make the equivalent of Contrex by putting minerals into distilled water.'

Production: 850 million litres (225 million US gallons) in 1993.

Exports: mainly to Belgium, the Netherlands and Germany.

Bottling company:
Société Générale de Grandes Sources des Eaux Minérales Françaises,
88140 Contrexéville,
France.
Owned: Nestlé Sources International.

Analysis	*mg/l*
Calcium	*467*
Magnesium	*84*
Sodium	*7*
Potassium	*4*
Sulphates	*1,192*
Bicarbonates	*384*
Chlorides	*7*
Nitrates	*2*

Evian: still, low mineralization.

ÉVIAN

Évian is France's best-selling still mineral water; its low mineral content makes it suitable as a family table water and also for mixing baby formulas. Évian water is also marketed as a facial spray in atomizers.

The benefits of the water from the spring, above the town of Évian on Lake Geneva, were first recognized in 1789, the year of the French Revolution, by the Comte de Lesser. The pure water, filtering down from the Alpine foothills, eased his kidney stones. By 1815 thermal baths were established close to what was known as Source Cachat, named after the family which originally owned the spring. A grand hotel was built to house the *curistes* coming to take the waters. In 1857 King Victor Emmanual of Savoy, in whose province Évian lay, authorized the establishment of Société des Eaux Minérales d'Évian. Soon Source Cachat water was being distributed to nearby Geneva and then to Paris in big glass *bonbonnes* packed in straw. Once the railway reached Évian in 1890, the waters began to be widely distributed. Exports to the United States started in 1905, and exports to Cuba alone soon reached nearly a million bottles a year. The Évian spring was declared *d'intérêt public* in 1926. The town of Évian-les-Bains, as it became, has benefited from this success through a judicious royalty of 1.5 centimes on every bottle sold. The revenue now earned on the annual sale of over 1.3 billion bottles (including the successful 'multi-buy packs' of various sizes) means

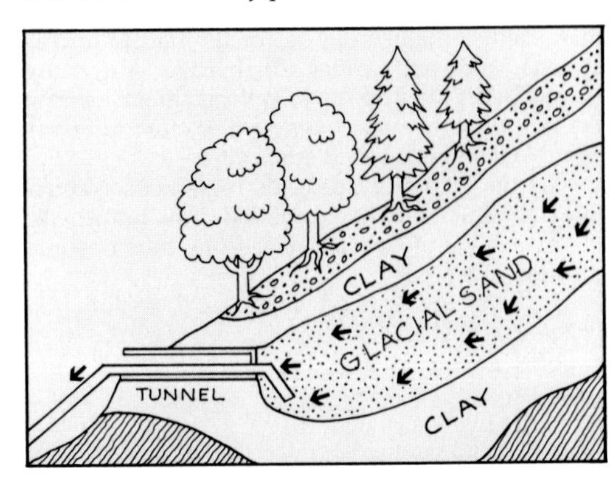

that the town of scarcely 6,000 people has little need of other local taxes and can afford to employ 100 gardeners, who earn it the accolade of the most flower-filled community in France.

The simple purity of Évian's water derives from the rain and snow on the Vinzier plateau above the town in the foothills of the Alps, which percolate slowly through gently sloping glacial sands for 15 or more years. Ultimately the water is trapped in sandy beds between two layers of clay. This natural reservoir, sealed off by the clay from any pollution from more recently fallen rains, is tapped by stainless steel drains at the end of a horizontal tunnel cut nearly 70 metres (230 feet) into the hillside above the town. It flows at a constant 11.6°C (52.88°F).

Évian is authorized to draw off 1,600 litres (422 US gallons) a minute of this natural flow, which is channelled to its highly automated factory 8 kilometres (5 miles) away. There it is bottled in conditions of exceptional hygiene, with visitors viewing only from a sealed gallery. The entire bottling area is slightly pressurized to keep out germs. Workers who have colds, mild infections or cuts are not allowed inside, but spend their shift instead tending the immaculate lawns and garden around the plant. Évian's constant concern for the purity of its water was best summed up by an executive who said: 'The water is our capital.'

Production: 1,285 million litres (339 million US gallons).

Exports: 37 per cent of production all over the world.

Bottling company:
SA des Eaux Minérales d'Évian,
22 Avenue des Sources,
74503 Évian,
France.
Owned: Danone (BSN)

Analysis	*mg/l*
Calcium	*78.0*
Magnesium	*24.0*
Potassium	*1.0*
Sodium	*5.0*
Bicarbonates	*357.0*
Sulphates	*10.0*
Chlorides	*4.5*
Nitrates	*3.8*
Silica	*13.5*

Tunnel leading to Évian spring is protected by strict security doors.

Visitors sample Évian water at the public fountain.

Perrier: naturally moderately carbonated, low mineralization.

PERRIER

Perrier is the world's best-known sparkling mineral water. 'I'll have a Perrier' has become an acceptable request, without endangering the party spirit. In France itself Perrier has over 30 per cent of sparkling sales.

Although Hannibal reputedly refreshed himself at Les Bouillens, the bubbling pool at Vergèze among the vineyards of the Languedoc plain, on his way to attack Rome in 218 BC, the real story of Perrier is shorter. In 1863 Emperor Napoleon III granted Dr. Alphonse Granier, a former mayor of Vergèze, the authorization to exploit the spring. In 1894 Dr. Louis Perrier from Nîmes leased the spring and began to market its water. Soon he found an ally in St John Harmsworth, the younger brother of Lord Northcliffe, founder of the London *Daily Mail*, who tasted the waters while on vacation. St John Harmsworth promptly sold his shares in the family publishing business to buy the naturally sparkling spring, and he devoted the rest of his life to building its reputation. Compagnie de la Source Perrier, with complete English management, was formed in 1906. Harmsworth was not deterred by an early car accident, which left him paralysed from the waist down. Indeed, he got the idea for the shape of the now familiar Perrier bottles from the Indian clubs with which he exercised to regain strength. Under his guidance, Perrier built up sales of 18 million bottles, almost half for export to remote corners of the British Empire where local water was too contaminated to mix with the whisky. Harmsworth died in 1933 and the Perrier bottling company's fortunes languished until 1947, when an energetic Paris stockbroker, Gustave Leven, took it back under French control. Over the next 45 years his initiative made the Perrier group, which he first expanded with Contrex in 1954, the world's largest producers and distributors of bottled waters. Perrier and its other waters were bought by Nestlé in 1992 to form the core of Nestlé Sources International.

Perrier water results from the natural mixing underground of three waters: one percolates through a gravel aquifer beneath the Languedoc plain, the second trickles down through natural faults in limestone from the Garrigues hills north of Nîmes, and finally a more highly mineralized hot water, of deep volcanic origin and laden with CO_2, pushes up through faults in the substrata. This cocktail of waters comes together in beds of sand and gravel, rich in silicon, between 11 metres (30 feet) and 28 metres (80 feet) down. The reservoir is sealed above from surface pollution by a 4.5-metre (15-foot) layer of impermeable clay. Centuries ago the water broke through a weak point in the clay to form the pool Les Bouillens, where people bathed

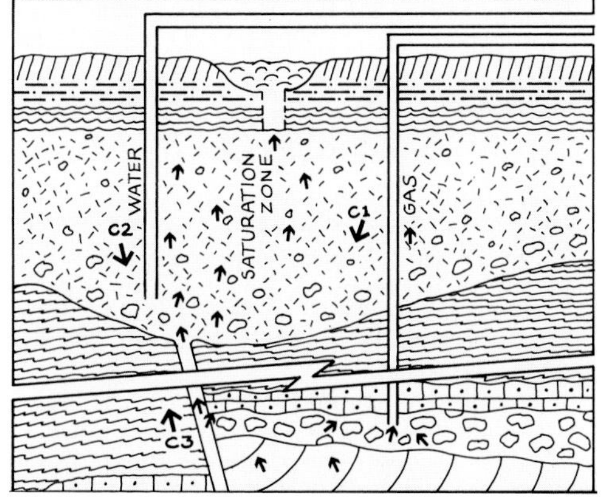

or collected water directly. The pool was filled in, however, early in this century. Perrier now obtains its water, which comes out at a constant 15.6°C (60°F), through stainless steel tubing drilled directly down over 80 metres (262 feet) into the underground reservoir, while the natural carbon gas is collected by drilling separately into the rock strata below. In the bottling plant water and gas are reincorporated in exactly the original quantities of 3.5 grams of gas per litre of water.
Production: 332.5 million litres (87.8 million US gallons).
Exports: 40 per cent of production.

Bottling company:
Société Générale de Grandes Sources des Eaux Minérales Françaises,
30310 Vergèze,
France.
Owned: Nestlé Sources International.

Analysis:	*mg/l*
Calcium	*145.0*
Magnesium	*3.5*
Sodium	*13.8*
Potassium	*1.1*
Bicarbonates	*336.7*
Sulphates	*51.4*
Chlorides	*30.9*

A kiosk in the gardens covers capped source.

A villa built by British press heir St John Harmsworth alongside Perrier source.

Vichy Celestins: naturally carbonated, high mineralization.

EAU MINERALE NATURELLE
DECANTEE REGAZEIFIEE AVEC SON PROPRE GAZ
VICHY (ALLIER)
VICHY
CÉLESTINS
VICHY ETAT
1L e
A CONSOMMER DE PRÉFÉRENCE AVANT 1988

VICHY CELESTINS

A glass of Vichy water anywhere in the world instantly soothes the average Frenchman suffering from homesickness.

To walk through the parks of Vichy, protected by wrought iron and glass awnings over the main pathways, to admire the jewel-like theatre in the Casino, a miniature Paris Opera built in 1865, and then to sip the waters of Vichy is to be transported back into another more elegant world: the France of Napoleon III, when the Emperor and his court and some 300,000 visitors would descend on Vichy each summer to rectify their health. Though many of the *beau monde* of Europe who came to Vichy in its heyday (there are streets named after Russian, Italian and English visitors) may largely have been suffering from overeating and its side-effects, others who came later were French colonial administrators who had wrecked their health in the tropics and sought to recuperate by drinking the waters each morning and evening in the conservatory-like Hall des Sources.

Citizens drink at fountain of Source Celestins.

Today both the Second Empire and France's colonies have disappeared, and with them much of the tradition that brought visitors to Vichy. But the Hall des Sources is still impressive. In one corner a glass dome covers the bubbling hot springs of Source Chomel, surrounded by layers of masonry stretching back in date from the 19th century to Roman foundations. Under the graceful roof, *curistes*, those drinking the waters under medical supervision, come and go, measuring their water doses, taken from the *gryphons*, in marked glass mugs. A mere 20,000 a year come today for a full three-week cure to sip the six Vichy waters (Hôpital, Chomel and Grande Grille – the three hot springs, and Celestins, Parc and Lucas – the cool waters) to prevent or heal kidney stones and intestinal disorders, or rheumatism and muscular ailments thought by French doctors to be soothed by water treatments.

But Vichy is beginning to see a new kind of visitor. Medicine has now come full circle and 'alternative' methods, rather than antibiotic doses, are growing again in public esteem. Stress ailments have replaced earlier illnesses. The Institut de Vichy has opened up a splendid wing of the old thermal station for a 10-day, slimming, massage and water-treatment package including baths of mineral water and local plant extracts. The French water, health and beauty traditions have mingled in a new formula, which modern Frenchmen and especially women, are quick to appreciate. A 10-day stay at Vichy, say the new ads, will add another 10 years to your life.

Vichy and its waters belong to the French state, and the springs and cultural establishments are run under concession by the Compagnie Fermière de Vichy, now part of the Castel group. The group is currently engaged in a thoroughgoing programme of restoration, to bring Vichy back to its former glory. Already superbly renovated are the Pavillon de la Source Celestins, a graceful colonnade around the spring where visitors can drink from the fountain; the Hôtel Pavillon Sévigné, a luxury hotel centred on the 17th-century house of the Marquise de Sévigné; and the café-restaurant in the park, looking like a clip from the film *Gigi*.

The only water that is bottled from among the many that emerge in Vichy is Vichy Celestins, high in bicarbonate of soda, and emerging at the coolest temperature, 17.3°C (63°F). Though Vichy's sales in France are moderate, 40 per cent of production is exported to French communities around the world for whom a glass of Vichy water is the closest thing to home.

Production: 45 million litres (11.89 million US gallons).

Exports: 40 per cent of production.

Bottling company:
Compagnie Fermière de Vichy,
1 and 3 Avenue Eisenhower,
03200 Vichy,
France.
Owned: Castel group.

Analysis:	*mg/l*
Sodium	1,200.0
Calcium	100.0
Potassium	60.0
Magnesium	9.0
Bicarbonates	3,000.0
Chlorides	220.0
Sulphates	130.0

Sipping prescribed waters in the Hall des Sources, Vichy.

Saint-Yorre: naturally moderately carbonated, high mineralization.

SAINT-YORRE

Saint-Yorre is a powerfully flavoured mineral water, with a taste it takes some time to acquire. But in France it is a popular thirst-quencher and appreciated as an aid to digestion because of its high bicarbonate level.

The waters of the Vichy basin (not to be confused with the springs inside the nearby town of Vichy) represent a unique occurrence in France. The whole landscape along the river Allier in central France seems to be alive with bubbling, iron-laden springs. A natural spring, spot-lit under a glass dome, erupts obligingly in the reception area of the Saint-Yorre bottling plant. Nearby, seven little huts signal other bubbling springs, which have been capped and piped into the factory. Because more than 200 springs within the 15,000-hectare (20,000-acre) protected zone all produce water of such similar composition, French law has accepted that they are the same water. The usual insistence, that a mineral water must come from one spring and be

Vichy town and river valley boasts dozens of mineral springs.

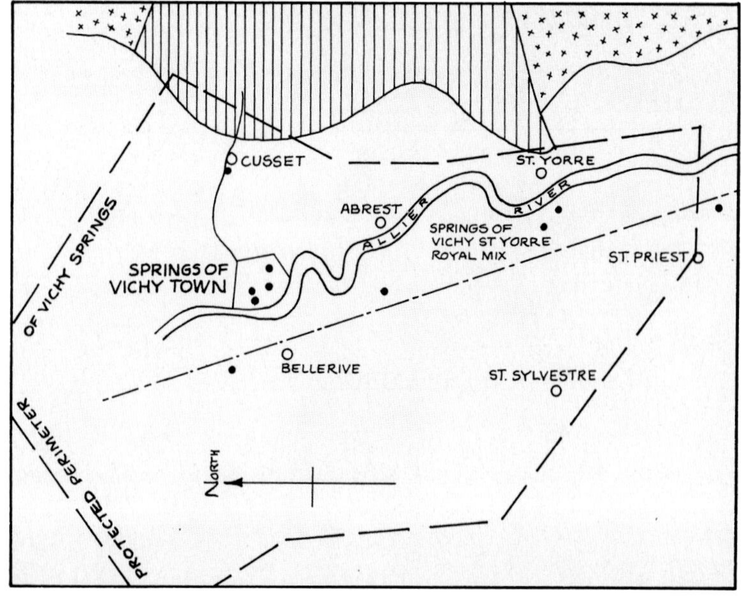

unmixed in any way, has been waived in the case of the allied waters of the Vichy basin. From its many springs, the Saint-Yorre company is permitted to mix the water for bottling purposes in what is called *la mélange royale* (the royal mix). Other areas of France that have attempted to market mixed waters have been denied a legal permit, and the unique character of the Vichy basin waters has been upheld in court.

The waters bubble up under carbon gas pressure from volcanic depths, and are, at origin, hot springs which cool off as they rise through fissures in the granite, acquiring a strong mineral content and numerous trace elements as they do so. The different springs have differing volumes of water behind them, and some on their own would permit only very low levels of production. But the Ministry of Mines has set an ample maximum limit of 200 million litres (52.8 million US gallons) a year to be taken from seven combined sources, supplying the plant at the small town of Saint-Yorre.

Discovered by a local pharmacist in the 19th century, the Saint-Yorre waters were at first bottled by hand from the springs with large speckles of red iron visible in their gassy midst. Today the springs are tapped by drills to maintain more sterile conditions. After the carbon gas has been extracted from the water and stocked separately, the iron is then removed by filtration. Once the filtration process is complete, the gas is then recombined with the water at the rate of 3.5 grams of CO_2 per litre. The highly automated factory at which this complex process is undertaken on a mass scale is run from a computerized control room on the first floor.

Production: 163 million bottles in 1993.
Exports: a mere 1 per cent of production.

Bottling company:
Société Commerciale d'Eaux Minérales du Bassin de Vichy,
03270 Saint-Yorre,
France.
Owned: Castel group.

Analysis:	*mg/l*
Sodium	*1,108.0*
Potassium	*80.0*
Calcium	*30.0*
Magnesium	*7.0*
Bicarbonates	*2,900.0*
Chlorides	*170.0*
Sulphates	*80.0*

Control room supervises mixture of waters and bottling process.

Vittel Grande Source: still, medium mineralization.

Vittel Bonne Source: still, light mineralization.

Vittel Hépar: still, high mineralization.

Vittelloise: artificially sparkling, light mineralization.

EAU MINERALE NATURELLE

Vittel

Grande Source

VITTEL

The town of Vittel on a wooded plateau in the Vosges has abundant springs with waters to suit all tastes. The best known is Vittel Grande Source, a still water with medium mineralization, widely sold in France, where it enjoys around 13 per cent of the still market. A more highly mineralized water, Hépar, which is rich in magnesium, also enjoys a small vogue in France, while Vittel Bonne Source, a lightly mineralized water, is marketed abroad. If you prefer sparkling water, then try Vittelloise, a still spring water to which carbon gas is added. Grande Source, however, is the flagship name.

The benefits of Vittel's waters were first noted by a colonial lawyer, Louis Bouloumié, in 1853, when he was cured of kidney problems by drinking its waters; he promptly bought the springs and set about building facilities for others to come and try the cure. The main buildings of the spa were built from 1880 onwards, including the great park and the graceful galleries where visitors can be seen regularly sipping the waters. Vittel's springs were declared *d'intérêt public* in 1903. The faded grandeur of the immense Grand Hôtel and the Casino has been rejuvenated by Club Méditerranée, the young-style French travel enterprise, which took over the hotel and sporting facilities in 1972. Today young couples leave their small children in a supervised Baby Club in the park, before cycling off for tennis, riding, jogging or swimming, with set hours to drink the waters.

Vittel waters boast a combination of calcium, magnesium and sulphates after their long journey up through layers of sandstone; these have a beneficial effect on the kidneys, liver and gall bladder. The company points to numerous tests showing that mineral water helps the body to eliminate more toxic substances than a diet of tap water. It maintains its sporting image by subsidizing sports events, while French Olympic teams have come to the town to train.

Although the Bouloumié family oversaw Vittel for generations, Nestlé, in an early foray into mineral waters, acquired a minority holding in 1969; this was stepped up to complete ownership in 1991, as part of the complete reshuffle of French waters which also brought Perrier under its control. The asset Vittel contributed to new Nestlé Sources

International was not just the export of its own waters, but extensive experience in exporting French taste, standards and know-how in tapping and bottling water. Vittel took a stake in the Lebanese spring Sohat, developed Baraka in Egypt, acquired Pisões-Moura in Portugal, and has provided technical advice on the development of waters in Kuwait and Abu Dhabi. Building on this international experience, an *Institut de l'Eau* was created at Vittel in 1993 to carry out research in mineral and spring waters for Nestlé Sources International.

Production: Vittel Grande Source: 700 million litres (185 million US Gallons); Hépar: 150 million litres (40 million US gallons); Vittelloise: 30 million litres (9.2 millions US gallons).

Bottling company:
Vittel SA,
BP 43,
Vittel 88805,
France.
Owned: Nestlé Sources International.

Vittel Grande Source:	*mg/l*
Calcium	202.0
Magnesium	36.0
Sodium	3.0
Bicarbonates	402.0
Sulphates	306.0

Vittel Bonne Source:	*mg/l*
Calcium	91.0
Magnesium	20.0
Sodium	7.0
Sulphates	105.0
Bicarbonates	258.0

Hépar:	*mg/l*
Calcium	575.0
Magnesium	118.0
Potassium	4.7
Sodium	12.5
Bicarbonates	376.8
Chlorides	7.0
Sulphates	1,548.0
Nitrates	1.2

Curistes sip Vittel water at fountain.

Galleries and parks surround Vittel springs.

Volvic: still, low mineralization.

VOLVIC

Volvic is a pure, light mountain water from the sparsely populated, volcanic highlands of the Auvergne in central France. Volvic is the fast-growing newcomer among the great names of French mineral waters. 'To understand Volvic, you have to see the Auvergne,' they say at the bottling company, and the remote, wild landscape, with its series of obvious volcanic craters, now carpeted with green and a sprinkling of forest, is beautiful and impressive. The springs and the bottling factory are in the heart of the National Park of the Volcanoes. Most of the volcanoes are not extinct, although there have been no eruptions in human memory. On our visit we were led 800 metres through a dark, wet tunnel to witness the tumbling underground torrent of the Source du Goulot, which is no longer used. The Clairvic spring, which is the one passed by the Ministry of Health in 1965 to be sold as a mineral water, can be seen in a small grotto within the grounds of the bottling factory.

The Source du Goulot with its underground torrent, was first discovered and capped in 1927 by Dr Moity, the mayor of Volvic. From 1938, the water from this and then from the Clairvic spring, were bottled and sold locally. After the Ministry of Health authorization, Volvic made the leap into the

Dr Moity, discoverer of the water.

Visitors peer at waters from underground mountain spring.

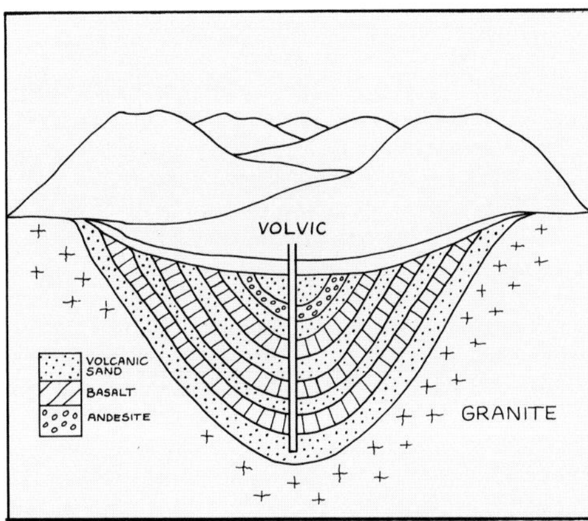

competitive national market in France, today selling 10 times its 1970 total. Its low sodium and calcium content, along with a fair magnesium level, makes it ideal for baby formulas and for those on low salt diets, while active in promoting good kidney function and useful in dermatology. Volvic's soft spring water even makes better coffee, it is claimed.

The waters spring from a huge natural filter in a shallow dip between volcanic hills, after draining for many years through porous basalt rocks on to a granite base. The deep source of the Volvic spring is some 80 metres (250 feet) below ground. The water rises at a cool 8°C (46°F) and should be drunk cool to enjoy its mountain flavour. The volume of water in the Volvic spring is virtually limitless, flowing at a generous 150 cubic metres an hour, allowing 100,000 litres (26,400 US gallons) to be bottled per hour.

The factory, while subject to all the usual checks and tests from the French system of inspection, institutes 300 checks a day at every stage of the bottling process, from the source to the moment the lid is fastened on the bottle. 'If people are buying pure mountain water, they have a right to the most stringent protection', says a Volvic director.

Production: 722 million litres (191 million US gallons).

Exports: 25 per cent of production to more than 60 countries, including Europe, the Middle East, Australia, the United States, the Far East and Japan.

Bottling company:
Société des Eaux de Volvic,
60 Boulevard Joffre,
92340 Bourg la Reine,
France.
Owned: Danone (BSN).

Analysis:	*mg/l*
Calcium	*9.9*
Magnesium	*6.1*
Sodium	*9.4*
Potassium	*5.7*
Bicarbonates	*65.3*
Chlorides	*8.4*
Sulphates	*6.9*
Nitrates	*6.3*

Authors prepare to venture 800 metres into mountainside to witness torrent.

REGIONAL WATERS

Although in France the famous mineral water names dominate more than in any other country, the regional water market is thriving and growing. Regional waters captured 38 per cent of still sales and 15 per cent of sparkling sales in 1993. Nearly 30 regional waters are now being bottled. They offer a tempting range of tastes from the bland Charrier, the least mineralized of all French waters, through the naturally sparkling Couzain-Brault of the Rhône-Alps region, or the Raphy-Saint-Simon source at Aix-les-Bains once renowned for its aphrodisiac qualities, to the powerful mineral cocktails of Rozana and the aptly named Geyser 4. Often less than a million bottles may be sold annually and encountered only in restaurants and supermarkets close to the source. But just as you might drink Côtes de Provence rosé wine on holiday on the Côte d'Azur, look out for Pioule still mineral water too; or in Brittany sample the waters of Plancoët. Although three of these waters are owned by the ubiquitous Nestlé Sources International (NSI), the majority preserve their independence. However, the Castel group, which now owns the waters at Vichy, also has a collection of eight other regional waters, thus securing a good niche in an expanding market.

Abatilles is a still mineral water from one of the deepest springs in France (at Arcachon, near Bordeaux), over 465 metres (1,500 feet) deep which is a sign of its freedom from all contamination. Owned: NSI.

Charrier water comes from the spring known as source Bonne-Fontaine, in the hills near La Prugne. Charrier is a still, cold water, 8°C (46.4°F), which comes from granite of the Bourbonnaise mountains, in wild, unspoilt country. Most French mineral waters are lightly mineralized in comparison with other European countries, but Charrier bears the distinction of being the least mineralized of all French waters, and in this sense is the purest. With a mere 36 mg/l of mineral salts, it is the next thing to distilled water, and indeed we met one Frenchman who used it to top up his car battery. Owned: Castel.

Chateauneuf comes from the Massif Central region, in the north-west Auvergne, an area of France known for its range of extinct volcanoes. The spa of Chateauneuf-les-Bains is still in use, and visitors who come to take the cure can drink three of its naturally sparkling waters. Only one, le Petit Rocher, is used for bottling, a water relatively highly mineralized with 1,800 mg/l of bicarbonates, calcium and sulphates. Owned: Castel.

Nessel, a lightly carbonated mineral water from the village of Soultzmatt near Strasbourg, holds a similar position in Alsace. The benefits of its waters were already being praised by a Franciscan monk named Tschmaser in 1272, and doctors in nearby Switzerland were writing about the waters of Soultzmatt throughout the 17th century. But it was

Louis Nessel, from a leading Alsace family, who really developed the water and gave his name to the source. He opened thermal baths in 1836, obtained permission to market the water in 1853, and in 1865 secured the *d'intérêt public* accolade. The water, from a reservoir over 61 metres (200 feet) down in the hills of Alsace, is very lightly carbonated, with just enough bubbles to coat the glass – making it *pétillante* rather than *gazeuse* – and is rich in bicarbonates.

Pioule. This moderately mineralized still water comes from an artesian source at Le Luc on the Côte d'Azur, on the broad plains bounded to the north by the foothills of the Alpes-Maritimes and to the south by the Massif des Maures. The spring was first discovered in 1880 when it was the only one in the locality to continue flowing through a 10-month drought. Pioule's water was declared *d'intérêt public* in 1884, and it enjoyed a brief vogue as a spa, seeking to rival Contrexéville. Today, its goal is more modest. 'We are the cheap local mineral water,' said the bottling plant manager, when we once called on a scorching day. 'People drink Pioule when they are thirsty.'

Plancoët is the mineral water known as the Sassay spring, which rises 20 kilometres (12 miles) from Dinard on the north coast of Brittany, from the granite rocks which overlook the small village of Plancoët. The only important water to be known in this region, Plancoët is valued for its exceptionally low mineral content. Owned: NSI.

Reine des Basaltes, in the beautiful hills of the Ardèche, is a naturally sparkling mineral water which springs from basalt rocks at the foot of the old Aizac volcano. It surfaces, without any drilling being necessary, at the rate of 10 litres (2.6 US gallons) a minute. The very high iron content is partially removed by filtration, at which point the natural carbon gas is stocked and reincorporated again later in the bottling process. Recognized as a mineral water by the French state in 1876, it has always been recommended for diabetes and liver problems, and as a diuretic.

Rozana was known to the Romans, and 19th-century archaeologists uncovered remains of Roman thermal baths near the present source. The water rises at Rouzat in the Puy-de-Dome region of central France, a formerly volcanic area. The spring is hot, rising at 30°C (86°F), and lightly carbonated so that a mere sparkle coats a wine glass. Recognized by the French state in 1932 and given the description *declarée d'intérêt public* the next year, the water is highly mineralized with nearly 3,000 mg of mineral salts per litre.

Thonon, from the Source de la Versoie on the shores of Geneva, like its famous neighbour Évian, offers a lightly mineralized still water filtered down from the Alps. Known to the Romans, and praised in the writings of St. François de Sales in the 16th century, Thonon was declared *d'intérêt public* by the French state in 1864. The water is cold (11°C, 51°F) at the spring, with a mere 354 mg of mineral salts per litre. Owned: Castel.

ITALY
INTRODUCTION

In Italy mineral water is accepted as an everyday basic necessity. In times of crisis, be it a strike or earthquake, the population runs out and stocks up on bread, wine and mineral water. If Omar Khayyam had been an Italian, he would have proposed a loaf of bread, a jug of wine and a litre of *acqua minerale*. Thus it is no surprise to find that Italians are easily the world's largest imbibers of mineral waters at 125 litres each annually.

In fact, bread, wine and water are the first things brought to your table in an Italian *trattoria* or restaurant. Drinking mineral water has nothing to do with abstention, with going on a diet, with driving. In Italy there are no breathalyser tests for motorists, no licensing laws, and no prohibition on alcohol in any way. But Italians naturally drink mineral water together with their wine, and not as an alternative. This is a long established tradition. In ancient Rome, slaves always mixed the water and wine, before a dinner party, and the Roman populace 'cut' their rough country wine intake with water to this day. In a typical Roman restaurant, you are brought water and wine, but only one glass.

The ancient Romans started it all with their interest in good mineral water springs, their thermal baths, and their preoccupation with the ideal of *mens sana in corpore sano* (a healthy mind in a healthy body). And since classical times, Italian waters have been studied and their effects noted. Throughout the Middle Ages and Renaissance, local waters beneficial to health sprouted miraculous reputations: 80 per cent of Italian waters are named after a saint. A primitive early system of distribution, among priests and apothecaries, can be seen from records for consignments of mineral waters in Vatican accounts.

The whole of Italy, mainland and islands, is rich in mineral waters and every region has at least one commercially bottled water, usually more, with the greatest number near the formerly volcanic areas of the Alps and the Appennine mountains. But there are eight mineral waters in Sicily and 12 on the island of Sardinia. In all, Italy has over 240 brands of mineral water owned by 180 bottling companies. Many more waters are self-bottled by the local inhabitants because the terrain makes the springs in some way unsuitable for commercial development. Near the Appia Antica in Rome, for instance, there is a good mineral spring which cannot be exploited commercially, because it occurs in the middle of an important archaeological site and there could be no access for lorries nor space to build a bottling plant.

So valuable does Italy consider her mineral waters that they all belong to the Italian nation, as natural patrimony, and bottling companies are merely granted a concession from the state. Cures at thermal establishments can be prescribed, and paid for, by the state medical system.

Italian law has long covered the definition and supervision of waters for spas and for bottling in great detail. Original legislation in a law of 1919 set out that a water could be classified as a natural mineral water only after studies of its therapeutic and hygienic properties, rigorous clinical and pharmacological tests, as well as chemical analysis and tests at the source to determine purity. Although more recently Italy has also been complying fully with the standard European directive 777 on mineral waters, this was not finally enshrined in Italian law, replacing the 1919 regulations, until 1992. One Italian anomaly also remains. Mineral water sources here are considered as publicly owned property, each belonging to its region, and are treated like mining concessions. The volume of water that may be taken out has long been controlled by the Ministry of Mines, but the life of a concession varies from region to region. Renewal of a concession can also be a lengthy administrative process.

This has not prevented the Italian mineral water industry forging ahead at an astonishing rate with a 240 per cent increase in consumption in the decade from 1983, far more than elsewhere in Europe. It continues to rise: 116 litres per head in 1992, 125 litres in 1993, and it is forecast to reach 150 litres by the year 2000.

The Italians' love of mineral waters is highlighted by the fact that their consumption of beer is a modest 40 litres annually and of soft drinks a splash at 22 litres (the Germans down 147 litres of beer). Yet within Italy there are sharp contrasts of intake, reflecting the diverse economic conditions in different parts of Italy. Thus in the prosperous northern plains of Lombardy consumption is 159 litres per person; in the poor villages of Calabria in the south it is a mere 18 litres (although this certainly understates the quantity as many country people in the south fill up bottles from local public springs). Taste, however, is

Elegant marble drinking fountains in San Pellegrino spa.

Boario spa reached its heyday around 1910.

changing.

A decade ago the Italians drank primarily sparkling mineral waters; still water had scarcely 15 per cent of the market. By the mid-1990s still waters, now packaged in plastic and PET bottles, accounted for 40 per cent of sales.

Moreover, Italian mineral waters are the most serious competitors for the ubiquitous French overseas. Travelling for this new guide, we sipped Crodo Lisiel with our pasta by Sydney Harbour, San Pellegrino in Dubai, Hong Kong, New York, Toronto and Zurich, and Ferrarelle in London. Italian waters rank consistently third in US imports of bottled waters (after French and Canadian). Exports are over 300 million litres (78 million US gallons) annually. The Italians themselves see little need to drink imported waters; imports are only 60 million litres (15.7 million US gallons), almost entirely from France.

The rapid growth has made mineral waters big business in Italy, increasingly concentrating them, as in France, into conglomerates. The top eight bottling groups now control virtually 80 per cent of sales, while the top 10 waters themselves have 60 per cent of the market. The top selling still water is Levissima, with a 14.5 per cent share, followed by San Benedetto, Panna and Vera; the most popular sparkling water is the delicately effervescent Ferrarelle with a 16.8 per cent slice of that range, followed by the sparkling versions of San Benedetto, Vera and Boario. Combining still with sparkling sales, San Benedetto is league leader at 11 per cent, with Levissima, Vera and Ferrarelle in close pursuit.

Meanwhile a power game for the control of various waters has seen the French giants acquiring significant Italian stakes. Danone (formerly BSN), who own such famous French labels as Badoit and Évian, has taken over Italaquae SpA, which controls such well-known Italian waters as Boario, Ferrarelle, Acqua de Nepi and Santagata. Italaquae has around 15 per cent of all Italian sales.

Meanwhile Nestlé Sources International (the inheritors of the Perrier group) owns directly the waters of San Bernardo and Vera, while also having a 25 per cent stake in the San Pellegrino group, which includes Claudia and Panna. Nestlé also secured a new niche in 1994 with 28 per cent of the equity in a new holding company Compagnie Financière du Haut Rhin, formed by the alliance of the San Pellegrino group with the Garma group of Raoul Gardini and Guilio Malgara.

Garma boasts successful labels like Levissima and Recoaro and is already prime challenger to Italaquae/Danone for the biggest sales. This new network gets Nestlé right into the heart of the Italian industry. But with Italians as the world's top mineral water drinkers, the battle royal is understandable. •

The contest to slake Italy's thirst is such that investment in new plant and machinery in the mineral water business is growing faster than similar capital spending in either goods or manufacturing. The prospect is bright. The market

Sangemini celebrated its centenary as a healthy water in 1986.

research group Canadean in its annual report on the Italian industry, *Acque Minerali Italiane*, forecasts that natural mineral water production in that country will rise from 5.4 billion litres (1.4 billion US gallons) in 1990 to 9 billion litres (2.3 billion US gallons) by the end of this century.

No wonder the French are bidding for a piece of the action.

National Association:
Mineracqua,
Federazione delle Industrie delle Acque Minerali e delle Bevande Analcooliche,
Via delle Tre Madonne 12,
00197 Rome, Italy.

Claudia: naturally lightly carbonated and with added carbon gas, medium mineralization.

CLAUDIA

Acqua Claudia has been known and appreciated since the days of ancient Rome. Although Rome itself was rich in mineral water springs, the special qualities of Acqua Claudia drew the ancient Romans 45 kilometres (27 miles) to Anguillara Sabazia, to sample the subtle taste of this water which emerges with a slight hint of effervescence, from volcanic rocks at a constant temperature of 23.7°C (74.6°F).

Clear evidence of the Roman era at Claudia was unearthed in 1934. A large ancient Roman villa was discovered in the lush park surrounding the bottling factory. Excavations revealed complicated water storage systems and remains of a Roman

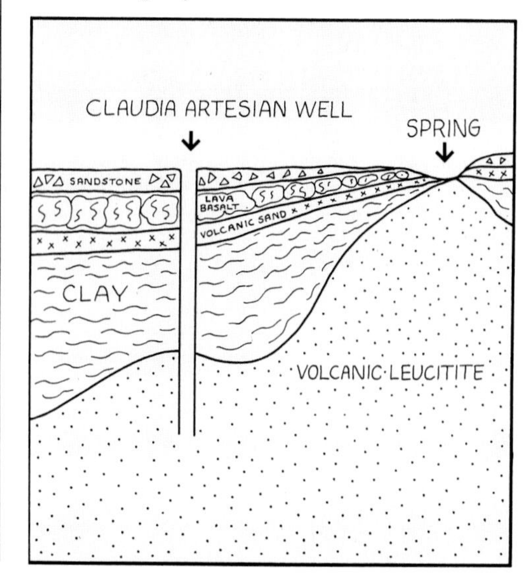

road linking the villa with the Via Cassia which leads directly to Rome. Lack of government funds has prevented the villa from being fully excavated, but what is known so far is tantalising. The villa dates from the first century before Christ and is a vast curved building 87 metres (285 feet) long.

Archaeologists cannot yet decide whether the villa was the private retreat of some wealthy patrician or an establishment where Romans came to take the waters in comfort. It might have had some religious significance since a depository for votive offerings was discovered. In the same way there is some mystery surrounding the name 'Claudia'. Who was the enigmatic Claudia who gave her name to the spring? A noble Roman matron, fabled courtesan or an austere priestess? Were Romans lured to face the arduous journey from Rome by thoughts of lavish hospitality or a desire for some ritual purification?

Acqua Claudia has been bottled since 1910 at a bottling factory in the medieval village of Anguillara on Lake Bracciano, a lake formed by the crater of an extinct volcano. After the San Pellegrino group bought the water in 1965, a new well was sunk to a depth of 34 metres (112 feet), and the water is pumped slowly to avoid losing the natural carbon gas valued for its flavour, and to prevent upsetting the water table. The water travels along stainless steel pipes to the bottling plant, where it is bottled in glass. About one-third of total production is bottled as it emerges from the spring with a low level of carbon gas, and despatched under a blue label. The other two-thirds has additional carbon gas and is marketed under a red label. The marketing strategy of the San Pellegrino group confines the distribution of Claudia to Italy, and to the region south of Orvieto (including Rome) in particular. A second spring was discovered 10 years ago which is sold under the label Acqua Guilia.
Production: 35 million litres (9.2 million US gallons).

Bottling company:
Claudia srl,
Via Pontina, 46.6 km,
Aprilia (LT),
Italy.
Owned: San Pellegrino.

Analysis:	*mg/l*
Sodium	*55.8*
Potassium	*78.2*
Calcium	*103.5*
Magnesium	*22.0*
Chlorides	*54.9*
Fluorides	*1.7*
Bicarbonates	*512.9*
Nitrates	*4.3*
Sulphates	*49.8*

Ancient lion fountain at site of Claudia spring.

Ferrarelle: naturally lightly carbonated, medium mineralization.

Acqua Minerale Effervescente Naturale

FERRARELLE

Ferrarelle's subtle touch of effervescence makes it one of the pleasantest of all Italian mineral waters to accompany good food and wine. And it is best appreciated if drunk at about 15°C (59°F), the temperature at which it comes out of the ground; if it is chilled it is not so beneficial to digestion. It is easily the most popular sparkling water in Italy, capturing almost 17 per cent of all sparkling sales (more than twice its nearest rival); and even though no still version is sold, it has 8 per cent of the total market. Such an asset has not been ignored in the power plays for Italian waters; once linked to the Sangemini group, it is now part of the Italaquae stable controlled by France's Danone. Indeed, Ferrarelle fits well into Danone's strategy of owning flagship waters.

Ferrarelle comes from the valley of the River Assano north east of Naples in the shadow of an extinct volcano named Roccamonfina. The active Vesuvius is also only 60 kilometres (37 miles) away. Pockets of carbon dioxide in the volcanic substrata give the water its natural carbonation. The water's reputation was already well established in Roman times. Hannibal is reputed to have rested his troops there before going on to sack the neighbouring village of Riardo during one of his forays against Rome. But that did not prevent Riardo later becoming a popular spa, conveniently situated close to the great Roman highways of Via Latina and Via Appia, linking Naples and Rome. Cicero and the naturalist, Pliny the Elder, both sang its praises. Pliny wrote, 'There is no greater miracle in nature than the waters of the valley of the Assano.' The Romans found the bicarbonated, calcium-rich water of Ferrarelle aided their digestion and eased their kidney stones. The fluoride content would also have helped their teeth. They also discovered, according to a local guide-book, that the water mingled excellently with wine, so that they could drink day and night without hangovers.

Ferrarelle's more modern reputation began in 1783 with a paper by a chemist named D'Andria praising its health-giving qualities. By the beginning of this century it was winning gold medals at exhibitions throughout Europe, and cultivating through opulent advertisements an

New test well is checked by director at Ferrarelle.

Riardo, the village of Ferrarelle Spring, just north of Naples.

image as *La Migliore Acqua de Tavola* (the best table water) to quaff in smart restaurants. At that time it was sold exclusively through the Istituto Nazionale Medico Farmacologico in Rome.

The Ferrarelle source is in a spacious park, filled with oak and lotus fruit trees, in open countryside still known as *Campagna Felix* for its gentle climate and fine fresh fruit and vegetables. The main spring has now been capped by drilling down to the aquifer 50 metres (164 feet) under the park, from which there is a natural flow of 1,500 litres (396 US gallons) per minute. A stainless steel pipe sprouting out of the grass in the midst of the park carries the water 200 metres (656 feet) to the bottling plant. Recently Ferrarelle has been drilling other test pipes into the aquifer; the water from these new outlets must have a constant flow and consistent analysis for two years before it can actually be bottled. And not all are suitable; we saw one emergence flowing freely that had such a high iron content that it had stained the surrounding grass dark brown. Since Italian regulations (unlike French and German) do not permit the iron to be filtered out, it cannot be used.

Like many carbonated waters, Ferrarelle separates out the carbon gas as it emerges from the ground, and later reinjects it on the bottling lines. For domestic sales it has always maintained the natural original proportion of gas (a light 2.5 grams per litre) just to give it that distinctive *pétillance*. But, when Ferrarelle was launched in the United States in 1979, it tripled the amount of gas in export bottles, feeling it had to match Perrier's greater sparkle. In practice the export drive was not a real success. So Ferrarelle still looks to Italy for 90 per cent of all sales; it is nationally distributed but is most widely encountered in restaurants or supermarkets south of Florence, and especially in Rome and Naples.

Production: over 600 million litres (160 million US gallons).

Exports: 5-10 per cent of production to United States, United Kingdom, Germany, France, Belgium and Switzerland.

Bottling company:
Italaquae SpA,
Via Appia Nuova 700,
00179 Rome (RM),
Owned: Italaquae/
Danone (BSN).

Analysis:	*mg/l*
Sodium	*50.0*
Potassium	*41.0*
Calcium	*408.0*
Magnesium	*23.0*
Chlorides	*21.0*
Fluorides	*0.6*
Nitrates	*4.0*
Bicarbonates	*1,513.0*
Sulphates	*5.0*

Panna: still and moderately carbonated, low mineralization.

PANNA

Acqua Panna, Italy's fifth best-seller, is a very light mineral water, equally suitable for making up milk formulas for babies' bottles or for the oldest members of the family.

The water comes from the Tuscan hills 40 kilometres (25 miles) from Florence. It emerges at a cool 8°C (46°F) at a height of 1,125 metres (3,700 feet) on the slopes of Mount Gazzaro, and then drops by gravity through stainless steel pipes to the bottling plant in the village of Panna below. The water is the lifeblood of the little community. All nine families work at the bottling plant, and many of them have done so for three generations, since bottling first started in the 1880s. In those days the water was bottled in demijohns and *fiaschi* – the straw-covered glass flasks now used for Chianti wine. These containers were then packed in straw and delivered by horse and cart. After the First World War, the water, always with carbon gas added, was put into ordinary glass bottles, sealed with a ball valve. In 1926 the bottles were corked, and in 1930 the present system of metal caps was introduced.

The concession to bottle the water changed hands several times and in 1937 it became part of an agricultural business run by Conte Bonicossi. This tradition is upheld today. Acqua Panna still produces, besides water, milk, meat and wine – a Chianti Classico: Casale dello Sparviero.

During the Second World War the empty glass bottles were hidden in the nearby lake to save them from destruction. A great deal of damage was done to the bottling plant by bombing. But with its high

Visitors view bottling plant from this sealed gallery.

mountain spring intact and its bottles, rescued from the lake, Panna was able to restart its operations soon after the war.

At that time 95 per cent of the water was carbonated. But after the San Pellegrino group took over Panna, it decided on a major change of emphasis. In 1970 it relaunched Panna as the first still mineral water in plastic bottles in Italy. Today over 80 per cent of Panna's sales are still. All told, the water has a 5.6 per cent share of the total Italian market. The plant itself has some of the most advanced technology in Italy, including one machine to check the plastic bottles for microscopic holes which would cause the cardboard packing cases to become damp and mouldy.

Panna's setting in the Tuscan hills is idyllic and the company owns many herds of dairy cattle and prime Angus/Chianina beef. Top management from the San Pellegrino group bring their guests to take part in shooting parties on the estate and, to make good the damage, their pheasant breeding farm raises about 10,000 pheasants a year.

Panna is still a family company. The present Director, Dr. Carminati, took over from his father, and some of the local families like the Baldini, the Biancalani and the Giuliani, have seen three generations employed at Acqua Panna. Eda Giuliani, for instance, whose father started work for Panna in 1904, runs the excellent *trattoria* at the bottling plant, which serves fresh beef from the company's herds. The *habitués* at the *trattoria* during the week are the truck drivers picking up loads of the water, which is distributed throughout Italy, but every Sunday the restaurant is packed with Florentines in search of traditional food, accompanied by the firm's Chianti Classico and Panna.

Production: 300 million litres (79.2 million US gallons).
Exports: 5 million litres (1.32 million US gallons) to Europe and the United States.

Bottling company:
Sorgenta Panna SpA,
Lungarno Vespucci 68,
Florence 50123,
Italy.
Owned: San Pellegrino.

Analysis:	*mg/l*
Bicarbonates	90.3
Calcium	14.7
Chlorides	10.0
Magnesium	5.3
Sodium	13.2
Sulphates	6.2

Restaurant on Panna estate serves home grown beef.

Sangemini: still, low mineralization.

Fabia: still, low mineralization.

Acqua Minerale Naturale

SANGEMINI

ITALIA (Umbria)

SANGEMINI

Sangemini is known throughout Italy for its purity and lightness, because of its bicarbonate minerals. It is always recommended for baby formulas and is to be seen tucked under the arm of patients entering hospitals and clinics.

The Sangemini water rises at 13°C (55°F) with a mere 0.98 grams per litre of carbon gas, on the western side of Mount Torre Maggiore in the olive and oak covered hills of Umbria, about 100 kilometres (62 miles) north of Rome. From ancient Roman times, when the town was called Casventum, a suburb of the Augustan city of Carsulae, the cold, sparkling waters were drunk to cure digestive ailments, and much appreciated, according to Tacitus, by the Roman army marching home.

The water was first analysed by Sebastiano Purgoti in 1837 and since then has become the most documented in Italian medical literature. Research into the rival digestibility of baby formulas mixed with Sangemini or distilled water have proved the 'live' mineral water to be very much easier to assimilate. Today, Sangemini is recommended not only for infants and nursing mothers, but for urinary complaints, for convalescence, after surgery, and in cases of stomach ulcers.

Visitors also come to Sangemini from May to September each year, to take the cure on the spot and drink the waters at the spa, the Stabilmento Idropinico, which is set in the midst of an extensive park, with many walks, tennis courts, and an open-air dance floor.

The restaurant, All'Antica Carsulae, at the entrance to the park owned by the San Gemini group, serves Umbrian specialities like pasta with the local black truffles. The company directors have realised the importance of preserving the mediaeval features of the town. The Abbazia di San Nicolo, dating from 1034, has been restored and the original portico, sold to the Metropolitan Museum, New York, in 1936, has been carefully copied and replaced.

Underground, Sangemini water filters through a fissure in calceous rock picking up carbon gas as it goes, and then is trapped above a thick layer of granite. It is protected from rainfall pollution by a further covering of granite 30 centimeters thick and a surface bank of gravel. However, in the late 19th century it was discovered that the waters of Sangemini were not entirely protected from intermingling with other waters in the same mountains, and the spring was capped inside the mountain.

In 1952 an entirely new system of capping and piping the water was introduced, complete with modern galleries leading into the hillside. The water is not pumped, but flows slowly through force of gravity down to the bottling plant in sterile and automatic conditions.

The original bottling plant was opened in 1889, when bottles were washed by hand and corked and sealed with lead. Eliso Spoldi, now nearly 90, remembers how the company offered a lunch to all workers who produced 100,000 bottles a month. Today, his son, Gianfranco, works in the near-by Acqua Fabia bottling plant which opened in 1974 and today bottles 60,000 litres an hour.

Next to the bottling plant is the glass factory which works 24 hours a day throughout the year turning out Sangemini bottles. Sangemini does not accept bottles back: each bottle is used only once, to avoid any risk of contamination.

Production: 150 million litres (38.7 million US gallons).

Exports: a mere 0.6 per cent of production.

Romans marched along Via Flaminia to drink water now bottled as Acqua Fabia.

ACQUA FABIA

This water was discovered by chance when the company was drilling nearby for fresh sources of Acqua Sangemini. Analysis proved that this was a different water. Fabia comes out at 13.3°C (56°F) from a depth of 65 metres (213 feet), and is a naturally still water. It has a beneficial action on the bile system, and it is good for the liver and kidneys. It is only bottled in its natural state.

Production started in 1974 of 1- and 0.5-litre foil-lined cartons. In 1982 it was also produced in plastic bottles, and in October 1984 individual plastic beakers were introduced to be used by the national airline, Alitalia, and other catering firms. Acqua Fabia is, above all, a water to drink when you are thirsty.

It is believed to be the water used in Carsulae, so as you quaff your way through the skies on Alitalia, you are drinking the same water enjoyed by ancient Roman soldiers on their long forced marches.

Production: 60 million litres (15.5 million US gallons) annually.

Exports: about 8 per cent of production.

Bottling company:
SpA dell'Acqua Minerale di San Gemini,
Via della Fonte,
05029 San Gemini,
Umbria, Italy.
Part of the Terme Demaniali di Acqui SpA group.

Sangemini:	*mg/l*
Sodium	*21.0*
Potassium	*3.8*
Calcium	*322.0*
Magnesium	*19.1*
Chlorides	*21.3*
Fluoride	*0.1*
Bicarbonates	*1,038.0*

Acqua Fabia:	*mg/l*
Sodium	*14.5*
Potassium	*1.4*
Calcium	*124.2*
Magnesium	*4.8*
Chlorides	*26.6*
Bicarbonates	*344.9*

San Pellegrino: moderately carbonated, medium mineralization.

Pracastello: moderately carbonated, medium mineralization.

SAN PELLEGRINO

World-wide, San Pellegrino is Italy's best-known sparkling mineral water to be found, almost as readily as pasta and Chianti, in Italian restaurants in 85 countries. It has been shipped to the United States for over 60 years. Curiously enough, it is not quite so widely found in Italy, where it ranks a modest fifth in sparkling sales. However, the overall San Pellegrino group, which includes seven other waters such as Claudia, Panna and Pracastello, and in which Nestlé Sources International has a 25 per cent holding, is third in the Italian league. Its position was further reinforced in 1994 by an alliance with the second largest group, Garma, through a holding company, Compagnie Financière du Haut Rhin (in which Nestlé also has 28 per cent). This new line-up of waters created the biggest network in Italy.

San Pellegrino's own water comes from three springs, with a combined flow of 50,000 litres (13,000 US gallons) per hour, at the foot of a dolomite crag towering above the river Brembana in the foothills of the Italian Alps near Bergamo, north east of Milan. The actual source is very deep, being 400 metres (1,300 feet) down through layers of limestone and volcanic rocks. Such depth accounts not only for the purity because of the long filtration period for rain and snow from the Alpine foothills, but also for the relatively high mineralization of over 1,000 parts per million. San Pellegrino is alkaline and emerges as a still water at a very warm 26°C (78.8°F). The local villagers and thousands of visitors who come for the cure every year drink it that way at a small public fountain, or at five fountains cut into the rock in the ornate drinking hall close by. The strong tang of mineralization lingers on the tongue when you drink it there, but that flavour is largely cancelled out by the artificial carbonation added at the bottling plant just down the hill, and slightly chilled on a restaurant table it is a refreshing, sparkling drink.

San Pellegrino's credentials as a 'magical water' benefiting digestive, liver and kidney ailments were established as far back as the 13th century. And later Leonardo da Vinci is supposed to have taken the waters. The first chemical analysis done in 1782 revealed San Pellegrino's special blend of mineral salts, including calcium, magnesium and chlorides, but a relatively low sodium (i.e. salt) content. Two hundred years later the analysis has hardly

changed, reflecting the consistency of the water.

The first thermal treatment facilities were opened in 1848, and commercial bottling began in 1899. A massive Grand Hotel was opened beside the river in 1904, while a new drinking hall on the hillside above was paved with marble and adorned with neo-classical murals. Alongside, the architect Romolo Squadrelli supervised the completion of the highly decorated art nouveau casino. For a while San Pellegrino was one of the most fashionable Italian spas. The faded splendour of that era is still preserved, but visitors today are the fitness conscious, not high society. San Pellegrino is seeking to cut a new image as a healthy holiday resort. Although hydrotherapy is available, and most people take a turn through the drinking hall, the emphasis is on sport and the open-air life.

Pracastello: While San Pellegrino is the renowned water from this valley, Pracastello, another spring on the same hillside above the village, has a slightly lower mineral content, and is piped into the same plant to be bottled, also with added carbonation.

Production: San Pellegrino over 220 million litres (60 million US gallons); Pracastello 38 million litres (10 million US gallons).

Export: San Pellegrino 60 million litres (15.8 million US gallons).

Bottling company:
San Pellegrino SpA,
Via Castelvetro 17/23,
20154 Milan, Italy.
Owned: Compagnie Financière du Haut Rhin (Garma/Nestlé).

San Pellegrino:	*mg/l*
Bicarbonates	222.65
Calcium	203.60
Chloride	73.80
Fluoride	0.60
Magnesium	56.90
Nitrates	0.80
Potassium	2.80
Sodium	46.50
Sulphates	540.00

Pracastello:	*mg/l*
Sodium	27.60
Potassium	2.80
Calcium	164.00
Chlorides	49.00
Magnesium	46.20
Nitrates	2.20
Sulphates	380.00
Bicarbonates	244.00

San Pellegrino's drinking hall was built for crowds.

Fiuggi: still, low mineralization.

FIUGGI

Fiuggi is known as 'the water of health' and has been famous throughout Italy since medieval times for its diuretic effect, and its action in breaking down kidney stones. Still, lightly mineralized, and radioactive, the water emerges from an aquifer 40 metres (130 feet) below the soil at a constant temperature of 11°C (51.8°F).

Three million visitors arrive during the season from April to November to drink the waters at Fiuggi about 64 kilometres (40 miles) south of Rome towards Monte Cassino. But its most famous admirers are from the past. The Pope Boniface VIII was near to dying from urinary infections and kidney stones until two priests made more than 50 journeys between 1299 and 1302 to bring him the waters of Fiuggi to restore his health. Perhaps even more renowned is the testimony of Michelangelo, the artist, who wrote in 1548, 'I am much better than I have been. Morning and evening I have been drinking the water from a spring about 40 miles from Rome, which breaks up the stone... I have had to lay in a supply at home and cannot drink or cook with anything else.'

The spa enjoyed its heyday after 1900, and after the opening of the Grand Hotel in 1910 became the summer drawing room of polite Roman society, and the location where many political declarations and treaties were signed. But after the Second World War, the hotels closed and fashion bypassed Fiuggi. In 1960 the water was leased by the present owners.

Today, one of the fountains at Fiuggi is named after Boniface, and this water is traditionally drunk each morning by visitors who gather in the modern drinking hall, which was completed in 1970 by architect Luigi Moretti. Here the reminder of the illustrious past is a few remaining neo-classical columns and ancient chestnut trees at the entrance to the park. The second fountain (though the same water) is the Anticolana, situated in colourful Italian gardens, and habitually visited each afternoon, when live entertainment is provided. Those who cannot come for the cure can buy the water locally, since Fiuggi is distributed nationally by Gruppo Italfin '80, which markets a wide selection of regional waters.

Production: 70-100 million litres (18-26.5 million US gallons).

Bottling company:
Ente Fiuggi SpA,
Via delle Betulle 1,
03014 Fiuggi (FR),
Italy
Owned: Gruppo Italfin '80.

Analysis	*mg/l*
Sodium	6.00
Potassium	4.60
Calcium	14.92
Magnesium	4.83
Chlorides	12.60
Fluorides	0.10
Sulphates	3.90
Nitrates	12.00
Bicarbonates	53.00

LEVISSIMA

Levissima is now the top selling still mineral water in Italy, capturing 14.5 per cent of sales in 1993, just ahead of its nearest rival San Benedetto. Since it is also sold as a sparkling water, it enjoys second place, with 9 per cent of the total Italian market, a position which, not surprisingly, attracted powerful suitors among mineral water giants struggling for more market share. The Garma group snapped up Levissima in 1992. This has brought Levissima firmly into the growing network of Garma, Nestlé and San Pellegrino orbiting through the Compagnie Financière du Haut Rhin.

What makes Levissima so popular? It is a very lightly mineralized water from the Valtellina mountains, near the ski resort of Bormio in the Italian Alps, which has caught the vogue for a healthy, refreshing drink after sport or on a hot day.

The low sodium content has been a considerable asset, while overall the dissolved minerals amount to scarcely 70 mg/l. The water originates from rock at 1,848 metres (6,160 feet) above sea level at a cool 8°C (46.4°F) and flows through glass-lined pipes down the mountain to the bottling plant in the valley below. While 80 per cent of the water is bottled in its natural state, two carbonated versions are also sold. (They secure Levissima eighth position in the sparkling league.) Levissima was recognised as a mineral water only in 1965. A small part of the sales are door-to-door in the locality, but it is now widely distributed throughout northern Italy.
Production: over 300 million litres (79 million US gals).

Levissima: still, moderately and highly carbonated, low mineralization.

Bottling company:
Crippa & Berger Fonti Levissima,
Via Algardi 4,
20148 Milan, Italy.
Owned: Garma.

Analysis:	*mg/l*
Sodium	*1.45*
Potassium	*1.75*
Calcium	*17.50*
Magnesium	*1.05*
Fluoride	*0.11*
Chlorides	*0.65*
Bicarbonates	*53.10*
Sulphates	*8.20*

SAN BENEDETTO

In the scramble to buy up best-selling Italian waters by giants like Danone and Nestlé, the best seller of all has so far eluded capture. San Benedetto, from the little town of Scorze just to the north east of Venice, enjoys the biggest sales in Italy. Actually the still and sparkling versions hold second place in each sector, but combined they secure first place with 11 per cent of the Italian market.

The secret is that at Scorze there are two abundant springs, whose combined production is 800 million litres (211 million US gallons). One of these springs, the original San Benedetto, has been known and appreciated for centuries. The beneficial effects of the water were chronicled in the *Annali della Republica Veneta* in 1700. This slightly alkaline water, with a nice touch of bicarbonates and modest amounts of calcium and magnesium, originates in the pre-Alps. It is bottled primarily as a still water, but 40 per cent of the production is carbonated. A second substantial source, Fonte Guizza, was discovered close by in 1965. This is a shade more richly mineralized, especially in calcium and sulphates, but still fits easily within the Italian *oligominerale*, or lightly mineralized, category. Just over half this spring's water is marketed as still, the rest with added carbonation.
Production: 800 million litres (211 million US gallons).

San Benedetto: still, lightly and moderately carbonated, low mineralization.

Bottling company:
Acqua Minerale
S. Benedetto SpA,
Viale Kennedy 65,
Scorze 30037,
Veneto,
Italy.

Analysis:	*mg/l*
Sodium	*7.6*
Potassium	*1.1*
Magnesium	*24.6*
Calcium	*42.9*
Hydrogen carbonate	*260.0*
Chloride	*2.0*
Sulphate	*5.2*
Silicate	*16.4*

Boario: still and moderately carbonated, low mineralization.

ACQUA MINERALE NATURALE

BOARIO

Boario Terme, a famous spa in the foothills of the Italian Alps north east of Milan, boasts a variety of springs, whose waters are used both for drinking and treatment. But it is the water of the Silia spring that has primarily been bottled since 1945, in still and carbonated versions, which together hold sixth place in the Italian water league. Boario is now part of the Italaquae group, owned by France's Danone.

Boario mineral waters have long been valued for their medicinal qualities, particularly in easing the digestion and preventing kidney stones. Tradition has it that the medieval apothecary Paracelsus made use of Boario water. In the 19th century, Italy's greatest novelist, Alessandro Manzoni, wrote many letters vowing he could only cope with his liver if he stuck to Boario.

In Manzoni's day (1845), the spring used was Antica Fonte, a simple fountain covered by a wooden canopy where local people filled demijohns with the water, and visitors stayed at a small inn while they took the cure. But by the turn of the century, Boario Terme boasted a splendid Grand Hotel and a pavilion decorated in the art nouveau style of Liberty of London. A railway link opened in 1908 made the spa newly accessible to visitors from Lombardy. An elegant drinking hall was built in the park in 1914. After the Second World War, the spa was modernized and enlarged, but traditional features like the Liberty Pavilion were carefully preserved. In 1955 the first of many prehistoric rock-carvings was discovered, and these have added interest for the many hundreds who visit the elegant spa each summer.

The old sources – Antica Fonte, Fausta and Igea – are now reserved for use in the thermal establishment, and another spring, Silia, is used for bottling. In practice, most bottling is from the Silia; 59 per cent is sold as still water, the balance is moderately carbonated (this version ranks fourth in sparkling sales). Almost half all sales come from door-to-door distribution, a somewhat rare occurrence in a nationally distributed water, although quite common with smaller regional springs.

Production: 270 million litres (71 million US gallons).

Fountain in the park of the elegant Boario spa.

Bottling company:
Italaquae SpA,
Via Appia Nuova 300,
00179 Rome,
Italy.
Owned: Danone (BSN).

Analysis:	*mg/l*
Sodium	*6.4*
Potassium	*1.8*
Calcium	*123.9*
Magnesium	*41.0*
Chlorides	*5.6*
Fluoride	*0.4*
Bicarbonates	*305.0*
Sulphates	*235.4*

CRODO VALLE D'ORO

Crodo Valle d'Oro: still and carbonated, medium mineralization.

Crodo Lisiel: still and carbonated, low mineralization.

Crodo, near the Swiss border in the Valle d'Ossola (Ossola means running water) produces two natural mineral waters, Valle d'Oro, known since the 19th century and recommended for dyspepsia, and Lisiel, discovered in 1956 and sold as a table water.

The waters which surface at Crodo, 13.7 kilometres (8 miles) from Domodossola on the Swiss border, are thought to come from Sempione and the Swiss Alps. They began to be studied seriously in the early 19th century, and Dr Luca Rossetti, director of Milan's largest hospital, analysed the waters and published a book about them in 1844. Crodo was too inaccessible ever to turn into a fashionable spa, although a few stalwart visitors did come to drink the waters. In 1900 grandiose plans for a Grand Hotel were drawn up but came to nothing. However, today, when most spas have experienced a decline, Crodo is developing as a resort offering healthy mountain walking as well as the local waters.

Societa Crodo was established and the waters first bottled in 1929, but the two sources used, Valle d'Oro and Cistella, both had a comparatively limited flow. The discovery of a new source, Lisiel, which has an abundant flow of 150,000 litres (39,600 US gallons) an hour, allowed real industrialization to take place. With a mineral content about one-tenth of Valle d'Oro, Crodo Lisiel is a light water, rising at the source at a constant 9.6°C (49°F).

The company bottles 25 per cent of its output with added carbon gas, and the remainder in its natural state. Valle d'Oro is bottled exclusively in glass, while Lisiel is also bottled in PET 500-ml and 1.5-litre bottles. Crodo is distributed nationally, though 40 per cent of sales are in north-west Italy. Societa Crodo is now part of the Dutch Bols group, and extensive modernization of the bottling plant has been carried out recently.

Production: 50 million litres (13.21 million US gallons) annually.

Exports: 8 per cent of production to the United States, the United Kingdom, Canada, Australia, Switzerland and other European countries.

Bottling company:
Terme di Crodo SpA,
Via Ludovico di Breme
25-27,
20156 Milan,
Italy.
Owned: Bols.

Crodo Valle d'Oro:	*mg/l*
Sodium	1.80
Potassium	5.50
Calcium	510.00
Magnesium	51.00
Chlorides	0.60
Sulphates	1,383.00
Bicarbonates	77.50

Crodo Lisiel:	*mg/l*
Sodium	5.70
Potassium	3.70
Calcium	60.00
Magnesium	1.70
Chlorides	1.70
Fluorides	0.16
Sulphates	104.00
Nitrates	3.20
Bicarbonates	103.09

Vera: still, lightly and moderately carbonated; low mineralization.

FRIZZANTE

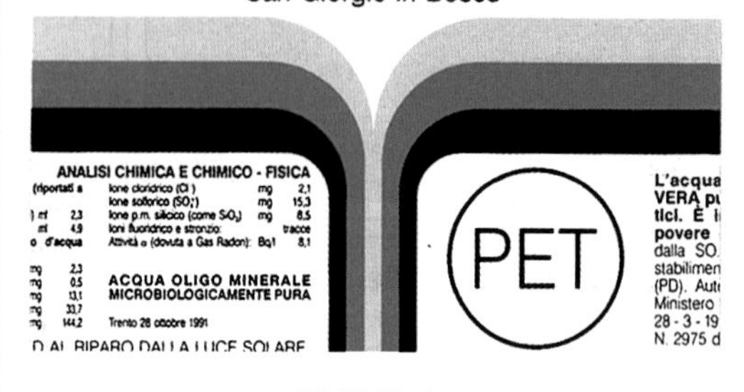

VERA

Although the San Giorgio spring at Bosco on the edge of the Dolomites between the little hill town of Bassano del Grappa (a great jewellery manufacturing centre) and Padua, has been known for centuries (the legend says it was a rallying point for warriors in Roman times, presumably to quench their thirst after battle), it is a relative newcomer as a bottled mineral water. Acqua Vera was launched only in 1979, but has made up for lost time. Catching the trend for lightly mineralized waters, especially in the cities of northern Italy, it has jumped into third place in both still and sparkling sales. Vera is the flagship water in Italy for Nestlé Sources International, which owns it directly, not through its holding in San Pellegrino.

The spring emerges, not far from the dashing torrent of the river Brenta, in the foothills below Bassano just before the plain that spreads out towards Padua and Venice. Water running off the mountains is filtered through gravel beds between layers of thick clay, and feeds into a series of aquifers at depths from 20 to 400 metres (66 to 1,312 feet). Acqua Vera comes from three aquifers between 20 and 60 metres; two other more highly mineralized waters Valviva and Sorgente Floresa are taken off the deeper aquifers. Vera itself has one of the mildest mineralizations of Italian waters, with total dissolved solids of a mere 162 mg/l. The sodium content is particularly low compared to most of Vera's Italian competitors for the light table water market, such as Fiuggi or Panna, so that it is acceptable to those on the most restricted diets. Vera is sold as still (60 per cent), slightly sparkling and sparkling mineral water, and is used in a selection of fruit-flavoured waters, such as blood orange, grapefruit, lemon and tropical. It is one of the few Italian waters distributed nationally.

Production: 600 million litres (158 million US gallons).

Bottling company:
SOGEAM SpA,
Via Valsugana 5,
35010 San Giorgio in Bosco,
Italy.
Owned: Nestlé Sources International.

Analysis	*mg/l*
Sodium	*2.3*
Potassium	*0.5*
Magnesium	*13.1*
Calcium	*33.7*
Bicarbonates	*144.2*
Chlorides	*2.1*
Sulphates	*15.3*
Silica	*8.5*

San Bernardo: still, lightly and moderately carbonated, low mineralizaton.

SAN BERNARDO

San Bernardo is a water with tradition and prestige. It comes from the mountains near the St Bernard Pass, and is especially popular in Piedmont, but also well-known in Lombardy and Tuscany. 'The Lightest Water in the World' was how San Bernardo was defined in the 1930s because of its balanced but low mineral content and its beneficial diuretic and digestive properties. San Bernardo is appreciated as an accompaniment to the great wines of Piedmont, and is light enough to be drunk by babies and invalids. The source rises 1,500 metres (4,900 feet) above sea level at a constant temperature of 6°C (42.8°F) in the maritime Alps near the border of the Piedmont and Liguria regions in north west Italy.

In the 19th century San Bernardo was appreciated by Napoleon, who spent some time in Garessio, and took the water for his bladder problem. Among the Italian royal family, which owned a hunting lodge in the area, King Victor Emmanuel was an enthusiast for the local water. In 1926 a royal decree authorised the 'Miraculous Fount of St Bernard' to bottle the water for the first time. A Grand Hotel was built to accommodate fashionable visitors taking the water. Today, the company maintains a small park where visitors can sit while trying the water.

Fonti San Bernardo has two plants which are near each other in the province of Cuneo. In the Garessio plant, which has been operating from early in the century, the water, is bottled in glass and PET, while in the high-tech Ormea plant, which was inaugurated in 1991, PET alone is used for bottling the water. Together the bottling plants employ 200 people, and produce around 250 million bottles (66 million US gallons). Three versions of the water are available – still, slightly sparkling and sparkling, and they are sold either in 25cc-, 50cc-, 75cc- or 1-litre glass bottles, or in 50cc- and 1.5-litre PET bottles. Most of the production is for the Italian market, in which San Bernardo holds a 5 per cent share.

Production: approximately 250 million bottles (66 million US gallons) a year.

Export: about 4 per cent of production, mainly to Belgium, the United Kingdom, Japan, the United States and the Caribbean.

San Bernardo, popular with former Italian Royal family, led poster design.

Bottling company:
Fonti San Bernardo,
Corso Galileo
Ferraris 26,
10121 Turin, Italy.

Analysis:	*mg/l*
Calcium	*12.00*
Magnesium	*0.60*
Sodium	*0.50*
Potassium	*0.31*
Bicarbonates	*36.60*
Sulphates	*2.00*
Chlorides	*0.60*

Acqua di Nepi: very lightly naturally carbonated and with added carbon gas, low mineralization.

ACQUA DI NEPI

This is a light, slightly effervescent mineral water from the countryside outside Rome, drunk with its own natural bubbles, which just coat the glass.

Acqua di Nepi, seen on some of the best tables in central Italy as an accompaniment to good food and wine, comes from one of the most famous Etruscan cities just 45 kilometres (28 miles) north of Rome. On the label it is stated that it comes from the original Roman spring which belonged to the Gracchi family, the pride of the ancient Roman Republic. Nepi itself for years stood out against the might of the ancient Roman Republic, but was defeated and added to Rome's dominions by Furius Camillus in 383 BC. The Emperor Augustus established a thermal spa and a 4,000-seat amphitheatre as part of the flourishing centre.

Today, Nepi's main claim to fame is still its water, which rises from a grotto cut 40 metres (130 feet) into the hillside, where it emerges at a constant temperature of 17.4°C (63°F) and with a vestige of natural sparkle. The water contains 1.53 grams of carbon gas per litre.

The water from the two sources – Antiche Terme dei Gracchi and Fonte Vivia – is piped to the bottling plant, situated in a natural hollow in underdeveloped countryside just outside Nepi. It is bottled exclusively in glass which best preserves its natural effervescence. Most of the water is bottled just with the natural fizz, but extra carbon gas is added to part of the production. Acqua di Nepi is now part of the Italaquae group, owned by France's Danone.

Production: about 50 million litres (13 million US gallons).

Bottling company:
SpA Terme di Nepi,
Localita Graciolo,
01036 Viterbo,
Owned: Italaquae/Danone (BSN).

Analysis:	*mg/l*
Sodium	*31.90*
Potassium	*43.00*
Magnesium	*25.70*
Calcium	*72.10*
Chlorides	*24.10*
Fluorides	*1.60*
Bicarbonates	*427.00*
Nitrates	*9.70*
Sulphates	*32.20*

THE WATERS OF ROME

The ancient Romans started it all. Wherever Roman legions came to rest, they made use of warm natural springs for baths and discovered mineral waters to drink alone or mix with wine. Most of all, they appreciated the springs of their own *patria*. Returning home, Roman soldiers drank at the spring still known as Laurentina from the laurel wreaths with which they were decked. Roman waters are naturally lightly sparkling, with a sharp almost lemony taste. Among waters appreciated today are:

Egeria, also known as Acqua Santa di Roma, rises close to the old Appian Way, near the famous tomb of Cecilia Metella, in an area rich in waters from the Alban hills. Volcanic rocks provide a marked mineral dose in the water which rises at a warm 18.6°C (64.5°F). The water was noted as *miracolosa* as an aid to digestion as early as 716 BC. Medieval Popes maintained its fountains and baths. Today it is bottled in its naturally sparkling state by the Mari family.

San Pietro is a naturally sparkling water which emerges at 16°C (60.8°F) near Marino between the old Appian Way and the new Via Appia. The fountain opens to the public at 4.30 each afternoon for a colourful crowd to fill demijohns and bottles at a bargain price. The bottling plant is owned by Gruppo Italfin '80.

Appia is valued for its light natural sparkle and bicarbonates which aid digestion. The spring, near the new Via Appia, first became popular with Mussolini's soldiers sent to Abyssinia in the 1930s; they topped up demijohns before embarking at nearby Ciampino military airbase. Now bottled for Gruppo Italfin '80.

Sao Paolo is a naturally effervescent water, rich in bicarbonates, which is appreciated as an aid to digestion. The water actually spurts from the spring under the force of its own carbon gas. It emerges at 20°C (68°F), with the characteristically sharp taste of all famous Roman waters, and is nicknamed *Fonte Acetosa* (Sharp Spring). The typical Roman today follows the ancient custom and uses this water to dilute the local white wine, which also has the characteristic flavour of the same volcanic soil.

The spring rises on the Via Casal di San Sisto, close to the Via Laurentina, the only Roman road which takes its name from the laurel wreaths worn by the returning victorious Roman army. Prehistoric tombs found nearby suggest that the water was known not only to the Romans, but to even earlier inhabitants.

Although Sao Paolo is very much a local city water, with a fountain open until 10 pm every night where people can fill their jars and bottles for a few *lire*, it is also bottled, mainly in glass, and distributed not just in Rome but throughout central Italy. Production is up to 20 million litres (5.2 million US gallons).

REGIONAL WATERS

Among the 240 bottled mineral waters of Italy, many are famous in their region, and are brought automatically to the table of visitors and Italians alike travelling through Tuscany, the Veneto or southern Calabria. Such local waters can be drunk very cheaply along with the local wines. Regional waters have increasingly been bought up by aspiring mineral water groups. Gruppo Italfin '80 controls 13 such springs; TSA Terme di S. Andrea has nine local sources, while in Sardinia six of the dozen springs are owned by Sarda Acque Minerali (SAM). Here are a few of the labels you may encounter in Italy.

Acqua della Madonna from Castellamare di Stabia, near Naples, in Catania. This still mineral water has only been bottled since 1962, but has been well known for centuries. The classical writer Pliny refers to its benefits, and that Spanish as well as Italian sailors took it with them on their ships when exploring the New World. Because ships regularly stopped near Naples to take on this water it used to be known as Acqua dei Navigatori. However, during the rebuilding of a local church in 1824, the prior managed to have the church boundaries extended to include the spring. As stories of miraculous cures effected by the water spread, the name was changed to Acqua della Madonna. In 1935, during Mussolini's conquest of Ethiopia, the water was sent to Massana on the Red Sea, and again in the Second World War was taken with the army to North Africa. Today it is found on many tables in southern Italy, bottled in its natural state, and with added carbonation.

Acqua dell'Imperatore, from Bolzano, in Trentino Alto Adige, near the Swiss, German and Austrian borders. This cool, still water, rich in calcium and bicarbonates, emerges at a constant temperature of 5.8°C (42.4°F) at Bagno Selvaggio, 1,333 metres (4,373 feet) up in the Dolomites. Though the bottling company is modern, and the water is sold today with light added carbonation, the spring has been famous for many centuries. Water jars from pre-Roman times have been found and the Romans themselves made a camp here named Littanum. Benedictine monks built a great basilica nearby at San Candido, where a small chapel is decorated with a triptych showing the saints curing the crowds with water.

Visitors came to take the cure from far afield, and when a professor from Innsbruck University wrote a treatise extolling the water in 1822, it became a centre of fashionable spa life and the Imperial family of Austria came for the cure. Though the spa was destroyed in the First World War, the label still boasts the German name Kaiserwasser.

Sorgente Flaminia, from Perugia, in Umbria. Nocera Umbra bottles three waters in Perugia, the Angelica spring, the Cacciatore spring, and Sorgente Flaminia, a moderately mineralized water which surfaces at a temperature of 12°C (53.6°F). The water was one of the earliest exported.

Fonte Napoleone, from Elba, Tuscany. This lightly mineralized water, bottled today with added carbon gas, emerges at a temperature of 11°C (51.8°F) at a height of 400 metres (1,312 feet) on Monte Capanne, on the island of Elba. Local tradition has it that Napoleon sweetened his exile in Elba by drinking this light water. The labels carry the Napoleonic eagle and Bonaparte bees as insignia.

Eureka from Lecce, in Puglia. There are very few mineral waters in this part of southern Italy, and it was with a sense of real discovery that Nicola Meleleo named the water he found in an olive grove in 1964. A bottling plant was constructed in 1971, and the lightly mineralized water is bottled in its natural state for therapeutic use and with added carbonation for table use.

Orianna, from Pesaro, near Urbino in Marche. This still water, rich in bicarbonates, which rises at a temperature of 11.5°C (52.7°F) is bottled at Cavignano Terme, and takes its name from the historic family, mentioned in Dante, who own the nearby castle. The water is also bottled with added carbonation.

Ciappazzi from Messina on the island of Sicily. This highly mineralized, still water was discovered in 1894 when a local highway was being constructed. Geologists believe that the water comes from the same fault in the earth's surface as the volcanoes of Etna, Vulcano and Stromboli. The main market for this still water, which bottles nearly 70 million litres (18.5 million US gallons), is Sicily and southern Italy. Owned: Gruppo Italfin '80.

Sandalia from Cagliari on the island of Sardinia, a highly mineralized, tangy water, is bottled still and with added carbonation, mainly for consumption on Sardinia. Owned: San Pellegrino.

Sorgente Alba from Valli del Pasubio in the Veneto is a lightly mineralized water, emerging at a temperature of 11°C (51.8°F). It is bottled still and with added carbonation. The water is distributed in Venice and northern Italy. Owned: Garma.

Santa Rita from Nè, near Benoa, in Liguria. A very lightly mineralized still water from Nè, near Chiavari and the beautiful gulf of Tigullio. The water rises at a temperature of 10°C (50°F) and is bottled in its natural state and in a carbonated version. Owned: Italfin Gruppo '80.

Fontenova from Citerno Taro, near Parma, in Emilia Romagna is a very lightly mineralized water, rising at a temperature of 12°C (53.6°F) from rocks at 447 metres (1,467 feet) north of Mount Zirone. It is bottled still, and with added carbonation.

Gajum from Canzo, in Lombardy is a light water which has been known for centuries. Writers such as Manzoni and Stendhal have written of their appreciation. It emerges at a temperature of 9°C (48.2F) and is bottled still and carbonated. Owned: Gruppo Italfin '80.

Ausonia from Terme di Bognanco in Piedmont. Ausonia is a naturally sparkling, moderately mineralized water, used for cures at the spa in Bognanco, and also as a pleasant table water. Owned: Gruppo Italfin '80.

GERMANY

INTRODUCTION

Germany is a land amply provided with thick forest and bubbling streams. From prehistoric times, and certainly under the Roman Empire, springs such as Fachingen and Gerolstein were centres of human habitation, no doubt chosen for their plentiful and healthful waters. In more recent centuries the tradition of taking a water cure at a local spa was as strong in Germany as anywhere, and always backed by the medical profession. The German *Bad* is nearly as well known as 'spa' in any language.

Today a large industry brings that same water to German families in bottles, either as a natural *Heilwasser* (health water), untreated in any way, or as mineral water, frequently with an added dose of carbon dioxide.

Germans now consume almost 100 litres (26.3 US gallons) each per year, the third largest intake in Europe after the Italians and the Belgians; though by contrast the Germans rank first in beer consumption (almost 150 litres) and in soft drinks. But the growth in mineral waters, as elsewhere in Europe, has easily outstripped other drinks, doubling in the decade between 1983 and 1992.

The growth has been helped, of course, by the re-unification of east and west Germany. While east Germany was under communist rule, although there are ample sources, minimal amounts of water were bottled according to strict state quotas. So when the Berlin Wall came down in 1989, per capita consumption was scarcely 5 litres, but within three years it had jumped to 35 litres and by the end of the century is expected to equal that in the rest of Germany. Satisfying that new taste has been big business for the well established west German bottlers. They not only took over the east German plants as they were privatized, investing heavily in new equipment, but started advertising extensively to stimulate sales.

Despite, or perhaps even because of this change, the German mineral water market – and its taste – remain in strong contrast to the rest of Europe. First and foremost is their love of fizz. While the trend in many European countries has been to still waters rather than gassy, all German waters are carbonated. Any still waters are imported (usually from France), but they constitute scarcely 3 per cent of the market. The Germans still prefer their waters to pack a punch. Indeed, until the mid-1980s, virtually all German waters were charged up with around 7 grams of carbon dioxide, a powerful cocktail known as *Sprudel*. And when we first called on Gerolsteiner Sprudel some years ago the marketing director told us their best-selling water contained 8.5 grams per litre of CO_2, but previously had a jolting 10 grams. However, as a slight concession to the trend elsewhere, less highly carbonated waters, with a more modest 2 to 4 grams per litre of CO_2, have since been introduced. These 'still' waters, as they are often referred to, now account for up to a quarter of sales.

This vogue has caught on, particularly in the

southern wine-growing areas, where white wine and water are drunk as a refreshing *spritzer*, which is more agreeable without too many bubbles. The addiction to fizz has also meant that, unique in Europe, virtually all water is still sold in glass bottles.

Another German characteristic is the regional, not to say local, nature of the market. Only two waters can truly be found everywhere in Germany: Apollinaris and Fachingen, and Apollinaris has been marketed world-wide for over a century. Gerolsteiner Sprudel and Überkinger are pushing to make national distribution true for some of their products also. But the managing director of Hassia-Luisen in Hesse was typical when he told us that his water sells throughout a district within a 160-kilometre (100-mile) radius of his factory. Franken

Tribute to royal patron, Kaiser Friedrich, who gave his name to his favourite water.

Brunnen near Nuremberg is content to dominate central Bavaria.

There seem to be two explanations for the regional nature of the market. First, the history of Germany, which has never had the national character of France, in which one city, Paris, always dominated all social life; in Germany separate princedoms went their own way. The second is the German consumer, who is among the most cost-conscious in the world. Since there is no resale price maintenance, the customer is charged what the market allows. If the German housewife can buy one mineral water at a few *pfennig* cheaper than another, she will opt for that. In a market which uses heavy glass bottles for all mineral waters and road transport everywhere, locally produced waters are bound to be cheaper. A good, cheap local water therefore can dismiss a more costly intruder on almost every occasion.

The diversity is such that around 400 mineral waters are bottled at 235 bottling factories owned by 150 companies. Although there has been some concentration of ownership over the last decade – there were 180 companies in the mid-1980s – and up to 50 companies sell between 100 and 800 million bottles a year, there are many still producing scarcely 10 million. The sources are concentrated mainly in the mountain regions of the Rhineland Palatinate, fairly close to the great Belgian source of Spa, while others proliferate in the Black Forest region of Baden-Wurttemberg from which, across the Rhine, it is not so far to the French sources of Vittel and Contrexéville.

Yet no German water has achieved the great slice of the market that those famous French names enjoy in their home territory. The best-seller, Gerolsteiner, has only a modest 8.8 per cent of the market, followed by Überkinger with 7.4 per cent, and Blaue Quellen (which embraces five different waters) with 6.8 per cent. Apollinaris, arguably the best-known name internationally, enjoys just 5.3 per cent. Indeed, taken together the top six selling companies corner under 40 per cent of sales, compared to over 60 per cent in France. The truly regional nature of the market is one reason why the international giants have largely kept out of Germany – except for Nestlé Sources International owning Blaue Quellen, and an alliance between Apollinaris and Schweppes.

That regionalism is also entrenched by the glass bottles; it is just too expensive to distribute them nationally. When we asked a leading French mineral water executive how he viewed the German market, he raised his eyebrows in despair. "Germany," he sighed, *"c'est un autre metier* (another business) to the rest of the world. It's stuck like that; you can't change it as long as you have those returnable glass bottles."

At least they all use the same bottle. Throughout Germany a communal bottle, tall and waisted, of bubbled glass, green for a *Heilwasser* or health water, and white for a mineral water, circulates. The same bottle, bearing many different waters, may be reused 40 to 50 times. The bottles

The bottling of Selters water around 1725.

are returned by supermarkets and restaurants to any manufacturer and one of the first sights on visiting a German bottling plant is of multitudes of bottles all with different brand labels scuttling along the assembly line for their first wash, de-labelling and anonymity. It takes about half an hour from this chaotic entry to the re-emergence of the same bottles filled, capped and bearing the brand name of the factory they are leaving.

This costly, but large investment, in automated bottling plants, especially by market leaders like Apollinaris, Gerolsteiner and Überkinger, has kept the Germans competitive with imported French plastic bottles. The communal bottle, however, may have solved some problems, but created others. No famous brand image bottle, like Perrier, exists. Little experiment or investment with PVC or PET bottles to match the French has taken place. Thus German waters have a hard time competing in long haul world markets, where shipping weighty glass is expensive. While bottlers such as Gerolsteiner Sprudel have developed a distinctive non-returnable bottle, close to 95 per cent of all waters sold in Germany are still in glass; what little is in plastic is mainly imported from France.

The tradition of glass bottles with plenty of fizz owes much to the original German regulations on mineral waters, which made a clear distinction between *Heilwässer* or health waters, drunk essentially for therapeutic reasons, and mineralized waters. The *Heilwasser* was regulated by the federal health authorities as a part of the drug industry and could not be tampered with in any way before bottling. Mineral waters came separately under food legislation and could make no therapeutic claim. The removal of iron or sulphur was permitted (which it would not have been with a *Heilwasser*), carbon gas could be added. But mineral waters had to contain at least 1,000 mg/l of dissolved minerals. Waters with less than 1,000 mg/l were called *Quelle*, spring, waters. The ubiquitous European directive 777, adopted in

Germany in 1984, changed all that.

Since then waters with less than 1,000 mg/l may be approved as mineral waters, provided they meet the other European criteria. One suspects that although the Germans bowed to the common European will, they did so reluctantly. The industry felt that 1,000 mg/l limit was an important criterion if a water was to qualify. And an official report observes that some aspects of the approval procedures were 'new and alien'. The concept of mineral water was strongly engraved on people's minds through a long tradition and they have not adapted easily to mass marketing of all kinds of waters in plastic containers. Hence the loyalty to glass. And it is worth observing that most of the individual waters discussed later in this section are in that tradition of highly mineralized waters. In short, they are drunk for what they do contain, rather than, as in America, what they do not. Moreover, all German mineral water bottlers also produce soft drinks made from their mineral waters.

The watchdog of this scene is the Fresenius Institute at Taunusstein, just north of Wiesbaden. Originally set up in 1848 by a Dr. Fresenius for the Duke of Nassau to undertake analysis of the waters of his domain, the Institute today makes regular assessments each quarter of most German waters. And its records going back over a century chart the virtually unchanged mineral content of many of them. Its reputation also attracts independent testing from waters abroad eager to prove their pedigree. The Institute, for instance, originally drew our attention to both Golden Eagle water in India and Canada Geese in Nova Scotia as waters worthy of including in this guide.

The present director, Dr. Wilhelm Schneider, is an acute observer of German waters, studying them as a professional and quaffing them as a thirsty sportsman. When we first got to know him, he at once pointed out a definite national taste for 'fitness waters with a salty taste like Kaiser Friedrich and a high sodium chloride and sodium bicarbonate level'. Dr. Schneider explained, 'If you sweat a lot when you are riding or hunting, you lose salt and you take it in again with a few glasses of Kaiser Friedrich. Some of our waters have a more alkaline taste, like Apollinaris or Fachingen, and counteract acid in the stomach. Others, like Gerolsteiner, contain a high calcium and magnesium level and have a hard taste that appeals on other occasions. Sulphate waters, like Neuselters, have a slightly bitter taste, which can supply another need. We have a long tradition of appreciating these different waters.'

Given those acquired tastes, it is easy to understand why German waters have had a hard time exporting. Production is close to 6.5 billion litres (1.7 billion US gallons) a year; exports a tiny splash at 35 million litres (9 million US gallons). Apollinaris, of course, has made its way around the world for over a century, Gerolsteiner Sprudel has tackled the British and US markets; Peterstaler makes a pitch in the US (but German waters rank a lowly sixth in American imports). Mostly German waters are to be found at home or just a short truck drive away in Belgium or the Netherlands.

However, the attitude is changing, not so much in direct export of German waters, but in the buying up or development of sources outside Germany in the way that the French have done. On an internal scale, this has already happened. Companies like Hassia-Luisen in what was West Germany have already been buying up sources in the former East German Republic. From there it is a natural step into eastern Europe.

Already Franken Brunnen has moved into Hungary, acquiring Apenta KFT in 1991. The latter bottles mineral water under the Apenta label (see below), and also diet and soft drinks. With German management and marketing skills and Hungarian technical know-how, Franken Brunnen reckons it is well-placed to capture a healthy slice of the growing eastern European market, where there is a lengthy tradition of mineral and *Heilwasser* (health water) consumption. Although the German company is reticent about its production figures, the Hungarian plant currently has an annual capacity of more than 6 million bottles of mineral water and 3 million of *Heilwasser*. Interestingly, with Apenta, Franken Brunnen has invested in another highly mineralized carbonated water. In a bold new move, private German investors have been involved in the development and launching of Canada Geese (see page 168), a highly mineralized spring in Nova Scotia.

But the most significant iniative has come from Gerolsteiner, already the best seller in Germany, which bought up the Italian group Terme di San Andrea in 1994. Terme di San Andrea has nine sources, which overnight provide Gerolsteiner with just under 5 per cent of the Italian market. Gerolsteiner, moreover, is now making much of its role of number three in Europe, after Nestlé and Danone. Is that an indication of things to come? Our impression is that in the next decade, although the domestic market may not change so much, the Germans will begin to challenge the French in foreign fields.

Fashionable drink in gay Nineties was whisky and Apollinaris.

National Association:
Verband Deutscher Mineralbrunnen,
Kennedyallee 28,
5300 Bonn 2,
Germany.

The label which appears on bottles of Hungarian Apenta mineral water, a moderately carbonated, highly mineralized brand, now owned by Franken Brunnen.

Apollinaris: highly carbonated, high mineralization.

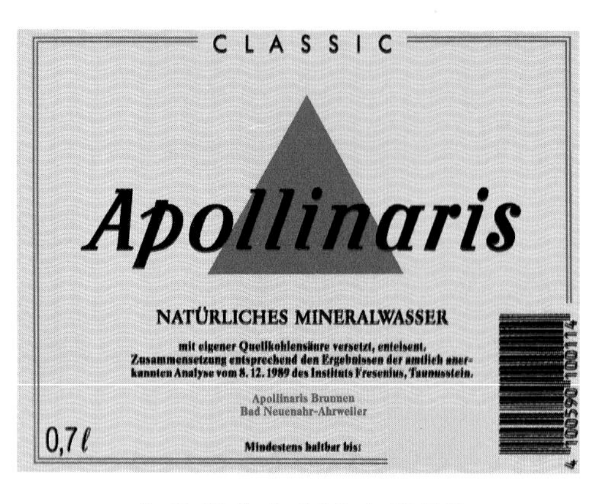

APOLLINARIS

The editor of the *British Medical Journal* first christened Apollinaris 'the Queen of Table Waters' nearly a century ago when it was much better known in England and the United States than at home in Germany. Today, however, this highly effervescent water with a very distinctive after-taste from its rich blend of mineral salts, is also the best known in Germany, and one of the few with national distribution.

The source, just outside the town of Bad Neuenahr in the Ahr valley, was discovered only in 1852, when a local farmer George Kreuzberg decided to find out why his vines refused to thrive on the terraced hillside. Digging down he soon found the soil strongly impregnated with CO_2 gas. A geologist at Bonn University told him there must be a mineral water spring below the surface. Sure enough, about 4 metres (13 feet) down Kreuzberg was rewarded with the discovery of a sparkling, clear water. It gushes lukewarm from deep volcanic origins through fissures in limestone. Later tests have shown that the water itself takes at least 30 years to filter down to the underground reservoir, and this long gestation accounts for the high mineralization.

Shipping Apollinaris on the Rhine in the 1870s.

He named the spring Apollinaris after St Apollinaris, a pupil of St Paul, who is buried nearby at Remagen, and has long been the patron of the valley. The next year he started selling the water in earthenware bottles. Its natural blend of calcium, magnesium, potassium and sodium, backed by its strong effervescent kick, soon earned it a reputation for easing digestive disorders. It came to the attention of Edward Hart, the editor of the *British Medical Journal* in London, who gave some to his friend, George Smith, who was so impressed that he approached Kreuzberg, asking if he would sell him a million bottles a year for Britain, and give him exclusive rights for exports elsewhere. Hart helped Smith with his advertising, coining the phrase 'Queen of Table Waters'. Apollinaris was an immediate hit in London, and Smith soon stepped up his order to 6 million bottles a year. The Prince of Wales (later Edward VII) liked what he called 'Polly' so well that a smaller 'split' bottle was introduced to please him.

By 1900, Apollinaris was selling more than 27 million bottles, of which over 25 million were for export; the Germans themselves hardly touched a drop. Indeed, with the exception of both world wars when exports halted, home consumption did not top exports until the mid-1950s.

Since then, the position has reversed. Apollinaris has ridden the rising tide of German mineral water consumption, pushing hard in its promotion the theme that it contains a unique blend of many of the minerals essential to the human body, which are often missing in modern diets. In 1991, Schweppes, the British soft drinks and mineral water manufacturer, took a 28 per cent holding in Apollinaris. The joint company is part of the empire of Brau und Brunnen, the brewery and drinks group.

Production: 375 million litres (99 million US gallons) for the group.

Exports: 14 million litres (3.7 million US gallons) to more than 60 countries.

Bottling company
Apollinaris & Schweppes GmbH & Co,
Postfach 10 01 51,
53479 Bad Neuenahr-Ahrweiler,
Germany.
Owned: Brau & Brunnen AG.

Analysis:	*mg/l*
Sodium	*425.0*
Magnesium	*104.0*
Calcium	*88.9*
Potassium	*25.2*
Bicarbonates	*1,580.0*
Chlorides	*137.0*
Sulphates	*112.0*

St Apollinaris is commemorated at the spring.

Neuselters: highly or moderately carbonated, medium mineralization.

Fürst Bismarck Quelle: highly or moderately carbonated, medium mineralization.

Rhenzer: highly or moderately carbonated, high mineralization.

BLAUE QUELLEN

Five mineral water springs, well spread out geographically, so that they can be distributed in the major cities of Germany, make up the 90-year-old Blaue Quellen group. Fürst Bismarck Quelle, named after the famous chancellor, is a spring located in the grounds of the von Bismarck family's Sachsenwald estate near Hamburg in the north of the country. Harzer Grauhof Brunnen was discovered in 1877 in the Harz forest near Goslar in north east Germany. The densely populated agglomerations in western Germany, the Cologne, Dusseldorf and Ruhr areas, are covered by Rhenser Mineralbrunnen, and the Rhine-Main region, with Frankfurt, Wiesbaden and Mainz, is served by Neuselters Mineralquelle located near Wetzlar in the beautiful Taunus hills. The southern cities of Munich and Stuttgart can easily be reached with the mineral waters from the Rietenauer Quellenpark.

All five mineral waters are available in three carbonation versions. Besides the medium type, with a carbonation of around 4 gm/l, Blaue Quellen has produced all brands without any carbonation for some years, but the traditional highly carbonated waters are still the major products.

The Blaue Quellen group ranks third in the German market with a share of almost 7 per cent.

Production: Blaue Quellen group: 670 million litres (177 million US gallons).

Fürst Bismarck Quelle: This is the only mineral water that may use the name of the great 19th century German chancellor, for it rises on land still owned by the Bismarck family. The water, which is quite lightly mineralized by comparison with many German waters, has been bottled since 1906. Today it is among the best sellers in such cities as Hamburg and Berlin.

Harzer Grauhof Brunnen: This spring was dis-

Herzer Grauhof: highly or moderately carbonated, low mineralization.

Kaiser Ruprecht: Heilwässer (health water).

Rietenauer Mineralquelle: carbonated, medium mineralization.

covered in the 19th century in the famous northern forest region of Harz, just south east of Hanover. The water has a low sodium content, so that it can be used for mixing baby formulas and by those on special diets. As a refreshing, light water of the forest, it is now one of the main brands found in Hanover, Bremen and Leipzig.

Rhenser Mineralbrunnen: The waters of Rhenser, near the Rhine, were first recognised in 1338 and are close to the town where Kaiser Ruprecht was made Emperor in 1400. The spring comes from a water table between 300 and 600 metres (1,000 to 2,000 feet) underground, and the source is now capped at 60 metres (197 feet). Commercial bottling began in 1872. Rhenser is a medium blend of mineral salts, rather rich in sodium (80 mg/l), calcium (118 mg/l) and bicarbonates (396 mg/l). A more potent nearby source is also bottled as a *Heilwasser* (health water), Kaiser Ruprecht, in honour of the early Emperor.

Neuselters: The spring was discovered in 1887 by Friedrich Wilhelm Neu, mayor of Selters, who was interested in hydrology. The water was named after him. He immediately started selling the water locally in stone bottles, and around 1920 moved over to glass. The spring is 172 metres (564 feet) below ground, and the water is pumped under pressure into the factory overhead. Neuselters is a balanced mineral water with bicarbonates, calcium, magnesium and sodium.

Rietenauer Mineralquelle: This spring, renowned since the 12th century, is in the village of Rietenau, north east of Stuttgart, which was long noted as a small thermal spa. The source was owned by the Marienthal convent for 400 years before becoming the property of the royal family of Wurttemberg. Commercial bottling began in 1960. The water has a good magnesium content (80 mg/l) and calcium (412 mg/l) content.

Bottling company:
Blaue Quellen,
Mineral und
Heilbrunen AG,
Brunnenstr. 2-8,
56321 Rhens.
Owned: Nestlé Sources International.

Fachingen (Heilwasser): very lightly naturally carbonated, medium mineralization.

FACHINGEN

Fachingen water, licensed as a *Heilwasser* (health water) is sold throughout Germany on two separate counts: first, it is the water you will be served in most restaurants in northern Germany if you ask for a bottle of still water to drink alongside your wine; and it is recognized as a curative water for a number of ailments by the German Federal Health Authority to be drunk at room temperature for maximum effect. The most expensive bottled water in Germany, Fachingen also retains a certain snob appeal.

The Fachingen spring rises in the remote, wooded (and, when we visited it, snow-covered) valley of the river Lahn, close to the village of Fachingen near Wiesbaden. Prehistoric communities knew the water, as finds of axes and implements indicate. Roman coins of the time of Varus testify to Roman occupation of Fachingen near the boundary walls built for protection from barbarian invasion. After the Romans were driven south, the spring was silted over and did not reappear until 1746 when it was uncovered by sailors pulling barges along the riverbank. From that point, Fachingen has become one of the most highly documented *Heilwässer* in German medical literature. It is widely used in German hospitals.

More romantically, Fachingen received a tribute from the German poet, Goethe, who wrote to his daughter-in-law Ottilie in 1843, 'The next four weeks are supposed to work wonders. For this purpose I hope to be favoured with Fachingen water and white wine, the one to liberate the genius the other to inspire it.' In more recent times, Rommel took Fachingen with him on his North Africa campaigns in the Second World War.

In strict conformity with Federal Health Authority rulings, no treatment of any kind may be applied to a natural *Heilwasser* in the bottling process, but it must be dispatched to the consumer in exactly the state in which it emerges from the ground. As Professor Wilhelm Pfannenstiel writes, 'Mineral and medicinal springs are individual phenomena of nature which are irreplaceable.' So the bottling plant has always been alongside the Fachingen well, and the water today is pumped up from a depth of 400 metres (1,300 feet) below ground by a pump placed 150 metres (490 feet) underground. The weight of water over the pump is greater than the pressure of CO_2 gas present in the water and the very small quantity of natural gas is therefore prevented from escaping.

In the 18th century Fachingen was bottled at the spring in stone jars and sealed with corks and dispatched to customers. In 1906, a well was sunk 11 metres (36 feet) down, and glass bottles were filled below, and then passed up the hill to the top floor of the factory, where a railway carriage was filled with cases of bottles. Even today, in walking round the new factory (constructed in 1973 and 1980) you come across a railway track and carriages disconcertingly positioned on the third floor. Fachingen still distributes some 10 per cent of its total production by rail, and 90 per cent by road.

Staatl Fachingen was managed for the state by a branch of the von Siemens family from 1894 to 1990, when it became part of the Überkinger group. It enjoys a position in Germany that is unique, successful in defiance of most of the laws of the market. Although its water is categorized as a *Heilwasser* rather than a mineral water, it sells in every restaurant. In cost-conscious Germany, it is expensive (around 1 DM a bottle) and unworried by the fact. In a sprawling country in which bottled waters have a strong local tradition, and basically sell only in their region, Fachingen is distributed and drunk nationally. It is the only water to travel by rail. In a country strongly attached to fizzy water, it is a still water, or what the French would call *perlé*; the small amount of natural CO_2 merely coats the glass.

Fachingen has had only one real problem, the volume of water. As sales grew steadily in the 1980s, it was rapidly coming up to the limit of water that could be taken off without disturbing the natural water table and mineral content of the water (a problem faced also by Badoit in France). The company looked for new springs, but found any nearby had a different, less subtle mineralized water. So it is on hold. Among other things this has limited its ability to export a unique water. Although now owned by Überkinger, it continues to trade separately.

Production: Over 100 million litres (26.3 million US gallons).

Bottling company:
Staatl Mineralbrunnen,
Siemens Erben,
Fachingen,
Lahn,
Germany.
Owned: Überkinger.

Analysis	*mg/l*
Sodium	*500.00*
Potassium	*14.60*
Magnesium	*62.30*
Calcium	*113.20*
Iron	*1.50*
Fluoride	*0.25*
Chlorides	*138.60*
Sulphates	*47.70*
Bicarbonates	*1,723.00*

Visitors to the mineral spring of Fachingen in 1834.

Gerolsteiner Sprudel: highly carbonated, high mineralization.

Gerolsteiner Stille Quelle: lightly carbonated, high mineralization.

St. Gero: Heilwasser, naturally lightly carbonated, high mineralization.

GEROLSTEINER

Gerolsteiner is not only the best-selling German water, but including the other waters the group now owns, such as **Bad Pyrmonter**, **Birresbomer Phonix** and **St Gero**, it ranks third in Europe behind Nestlé Sources International and Danone. The group has around 9 per cent of all German sales, and enhanced its international position in 1994 by purchasing the Italian group Terme di San Andrea which owns nine springs in Italy. That could be just the beginning of a challenge to the French giants. 'With both national and international acquisitions we are preparing for successful competition within the European market,' said Dr. Peter Traumann, the group's managing director.

The heart of this growing empire, however, remains the tiny town of Gerolstein, in the remote Eifel hills in Rheinland-Pfalz, which lies at the bottom of a valley 3 kilometres (1.8 miles) wide and 40 kilometres (25 miles) long, formed on the earth's crust 350 million years ago. The valley was the centre of warm seas, an ancient lagoon, with coral reef surrounds. The reefs and later volcanic action resulted in the cliffs which today dominate the town on both sides. Because of its long centuries of silting over the lagoon, Gerolstein has become a well-known centre for fossils of ancient land and sea creatures and for mineral waters. Gerolstein's water is a deep ancient water, in which the CO_2, formed from contact with volcanic magma, provides the thrust to push it up through layers of dolomite, where it picks up its strong mineral content. When the Gerolstein company taps new wells, they shoot 10 to 15 metres (35 to 50 feet) high before they are capped. The wells are rebored every 20 years, because the action of the CO_2 and water gradually destroys them.

The main Gerolstein spring emerges with 3.5 grams of carbon gas per litre, and more of its own gas is added in bottling. Gerolsteiner Stille Quelle sells with 4 grams of carbonation per litre, and Gerolsteiner Sprudel sells with 8 grams per litre, a resounding fizz. Gerolsteiner Sprudel has 8 per cent of the sparkling market and Gerolsteiner Stille Quelle has 11 per cent of low carbonated sales. Together 560 million litres (148 million US gallons) of these waters filled 823 million bottles in 1993.

The valley and its waters were known to prehistoric man, and to the Romans. Pots and coins from Roman times have been found surrounding Roman remains of artesian wells. Bottling started in the mid 19th century, when the railway was brought from Cologne to Trier, and it was possible to send stone bottles of the water to a wider world. The company moved over to glass in the 1930s and its product became a leading brand, but the factory and its production were left in ruins at the end of the Second World War.

Although a high proportion of sales from the three bottling plants in the town are to the nearby cities of Cologne, Dusseldorf and Dortmund, national distribution is achieved. As well as sharing the communal mineral water bottle of the whole German industry, the company has developed its own bottle for distribution to Berlin or Bavaria or for export where it is pointless to plan on returns. However, this accounts for only about 2.5 per cent of the total bottling.

The source, capped in 10 wells at a depth of between 80 and 180 metres (260 and 590 feet) is immediately under Factory Number One, and then piped some 3 kilometres (1.8 miles) to the other two factories further down the valley. The temperature of the mineral water is a constant 11°C (52°F). The water is sold with the slogan 'our success is the taste'; the Gerolstein spring, although high in magnesium, calcium and hydrogen carbonates, has very little sodium content so the flavour is not strong. The St. Gero Heilwasser, which is rich in calcium and magnesium and is very mildly naturally carbonated (a mere 1.9 g/l), comes from another local source from which it is piped with great care so as not to lose its small quantity of CO_2 before bottling.

In 1993, Gerolsteiner's export sales reached 72 million litres (19 million US gallons) 10 per cent of production dispatched to 25 countries.

Production: 720 million litres (190 million US gallons) for the group.

Exports: 72 million litres to 25 countries.

Bottling company:
Gerolsteiner Brunnen GmbH & Co.,
Brunnenstrasse 1,
D 54568 Gerolstein,
Germany.
Owned: 51% Bitburg Brauerei.

Gerolsteiner Sprudel:	mg/l
Sodium	128.20
Potassium	11.90
Magnesium	112.50
Calcium	363.70
Strontium	2.69
Fluorides	0.15
Chlorides	38.09
Sulphates	33.70
Bicarbonates	1,917.00

St Gero:	mg/l
Sodium	174.70
Potassium	14.10
Magnesium	120.50
Calcium	407.30
Strontium	3.49
Chlorides	72.30
Sulphates	37.90
Bicarbonates	2,161.00

Mineral water feeds the fountain in Gerolstein's park.

Hassia Sprudel: highly carbonated, medium mineralization.

Luisen Brunnen: highly carbonated, medium mineralization.

Bad Vilbeler Elisabeth Quelle: highly, moderately and lightly carbonated, medium mineralization.

HASSIA & LUISEN

If you are visiting Frankfurt or anywhere in the Lander of Hessen, then you will encounter the various waters of the Hassia & Luisen group bottled in the town of Bad Vilbel. The group is the market leader in that region and pushing for fifth place in the German league, but remains a family business in the fourth generation of the Hinkels who started it. They have four mineral waters, bottled with various degrees of carbonation: Hassia, Bad Vilbeler Elisabeth Quelle, Luisen Brunnen and Niederlichtenau.

Bad Vilbel is, indeed, a town of mineral waters, situated above a series of fissures descending into the water table of the ancient volcanic mountains of Vogelsberg, east of Frankfurt. Once Bad Vilbel boasted 22 mineral water companies; today it still has 13, following the mergers of some tiny sources.

The Hassia spring was discovered in 1864, when its water was first put into stone bottles and taken into Frankfurt by horse and cart. The picture, taken in 1905, shows Hassia bottles being loaded on carts for market. The small boy holding the bottles is the father of the present director, Günter Hinkel. The Hassia water is now tapped at 300 metres (980

Setting out to Frankfurt with the deliveries in 1905.

feet) and piped to the factory nearby. The original spring, however, can still be seen bubbling up under glass inside a small rose-covered temple in the nearby municipal gardens. The quantity of water is virtually limitless, so Hassia has been able to meet Germany's fast rising demand.

Hassia merged in 1982 with another local family business, Luisen, whose water had been bottled since 1882, to form the present group. Their panorama of waters is characteristic of the German taste. Traditionally, they had plenty of fizz. 'That's the way the cusomer likes it.' Gunter Hinkel told us when we first called on them. But he added significantly, 'Our "still" water now has the fastest growth in the group.'

So today you can take your pick of the old style, highly carbonated Hassia Sprudel, or the mildly carbonated Hassia Liecht. Luisen Brunnen is sold chiefly as a highly carbonated water, but another local spring, Bad Vilbeler Elisabeth Quelle is sold with the full range of highly carbonted, moderately carbonated and lightly *(stille)* carbonated water.

The abundant supplies from the Hassia source also enable the company to bottle a flavoured mineral water, Hassia Dry Lemon, and to use it for a selection of fruit juices, in common with many German bottlers who have ample sources.

After the dismantling of the Berlin Wall and the reunion with the former East Germany, Hassia-Luisen was one of the first bottlers to move into the nascent water business there. In 1990, the company took over the Neiderlichtenau mineral water source, near Brandenburg in Saxony, thus gaining an instant stake in this fast growing market. Niederlichtenau is now the leading brand in the former East German Republic, where consumption is up seven-fold in just four years.

Production: 200 million litres (53 million US gallons) for the group.

Bottling company:
Hassia & Luisen Mineralquellen
Bad Vilbel GmbH & Co.,
Giessener Strasse 18-28,
61118 Bad Vilbel
Germany.

Hassia source is capped under floral temple.

Hassia Sprudel:	*mg/l*
Sodium	232.0
Potassium	25.0
Magnesium	36.0
Calcium	176.0
Chloride	129.0
Sulphate	37.0
Bicarbonates	1,130.0

Überkinger: highly or moderately carbonated, high mineralization.

Die gesunde Art den Durst zu stillen

Überkinger

Natürliches Mineralwasser
enteisent und mit Kohlensäure versetzt.

Wohlschmeckend, bekömmlich, erfrischend und gesund.
Vorzüglich geeignet zum Mischen mit Wein und Fruchtsäften.
Berühmt seit dem 12. Jahrhundert.

Mineralbrunnen AG. Betrieb Bad Überkingen.

ÜBERKINGER

Wherever you travel in southern Germany, the red diamond on the label of the Überkinger group mineral water bottles will pop up on your restaurant table, in your hotel room and on supermarket shelves. Überkinger is the number one refresher in the area that stretches from the Black Forest in the west to Bavaria in the east. Überkinger and the four other mineral waters in the group are sold highly carbonated or in a moderately carbonated version as Überkinger Quelle, etc. The group also owns three *Heilwässer* (health waters). All told, the Überkinger group ranks second in Germany, with 7.4 per cent of sales.

The Überkinger water in the Alb mountains of Baden Württemberg was known in the Middle Ages, when visitors bathed in its hot springs and drank the cooler waters. From the end of the Thirty Years' War (around 1750), the drinking water was put into stone jars and sold in the locality and by 1870 a glass bottling plant and a spa with thermal treatments had developed. But it was only after 1950 that the bottling company at Bad Überkingen, armed with the communal bottle of all German mineral water producers (which can be returned to any bottler and used up to 40 times) and good modern roads, was able to take its water all over southern Germany and especially to the prosperous triangle of Stuttgart, Munich and Nuremberg.

Bad Überkingen is the site of four different springs which produce separate waters, the most obvious visually being the thermal waters rising from a depth of 300 metres (984 feet) at a temperature of 30°C (86°F). The steam from this water shrouds the park and hilltop temple dedicated to the poet Schiller in constant romantic mist. The water that is bottled as Adelheid Heilwasser comes from 60 to 70 metres (195 to 230 feet) deep and is a moderate 10 to 12°C (50 to 60°F), while the main spring of Überkinger water is relatively shallow at a mere 50 metres (165 feet) under the soil, and the same moderate temperature. Both waters for bottling are pumped from just beneath the factory into the production line. In both cases the waters are oxidized to remove iron, and CO_2 is removed at the same time. The waters are reinjected with CO_2 in the bottling process, but the CO_2 is imported frozen from elsewhere. Adelheidquelle has 2.5 grams of CO_2 per litre; Überkinger Quelle 3.4 grams per litre and Überkinger 8 grams per litre.

Since the Überkinger group also contains Teinacher from the Black Forest, Remstal Sprudel, Kisslegger and Imnauer Apollo, it is one of the leading bottlers in Germany. While the group has mainly been content to dominate its region, it has pushed for national distribution over the last decade. Initially it used one of its *Heilwässer*, Hirschquelle, as a pilot for national trials, seeking to challenge Fachingen in that specialised field. However, in 1990, it actually took over Fachingen, although that water continues to trade separately.

Despite 10 separate wells for tapping Überkinger waters, the rapid increase in sales through the 1980s led to the inevitable search for more abundant sources. In fact, Teinacher, in the Black Forest, has ample water and so has been pressed into greater prominence to help keep pace with German thirst. Linking its waters to sport and health, the Überkinger group has built spa hotels at Bad Überkingen and Bad Teinach. The hotels offer up-to-date thermal treatments in well designed, modern buildings. The resort at Bad Teinach (see picture) has proved particularly popular with health-conscious holiday-makers.

One interesting variation in German taste is noticeable in contrasting the sales of Überkinger with some of its northern equivalents. Southern Germany is not only thirstier than the north, consuming more than the average level of mineral water, but is also much more disposed to like moderately carbonated water as compared to the fizzy northern *Sprudels*. Since this is a wine-growing area, the frequently referred to correlation between wine and less carbonated waters seems to operate in southern Germany, which has more in common with France and Italy.

Production: 480 million litres (127 million US gallons).

Bottling company:
Mineralbrunnen
Überkinger-Teinach
AG,
7347 Bad Überkingen,
Germany.

Analysis:	*mg/l*
Sodium	1.180.00
Potassium	20.00
Calcium	26.05
Magnesium	17.02
Strontium	2.90
Lithium	1.05
Chloride	106.40
Fluoride	2.70
Sulphates	1.302.00
Bicarbonates	1.495.00

The new spa attracts visitors to Bad Teinach in the Black Forest.

Kaiser Friedrich Quelle: lightly, moderately or highly carbonated, high mineralization.

Rosbacher: highly or moderately carbonated, medium mineralization.

Kaiser Friedrich Quelle:	*mg/l*
Sodium	*1,419.00*
Potassium	*16.00*
Magnesium	*4.09*
Calcium	*4.79*
Chlorides	*795.50*
Sulphates	*336.10*
Bicarbonates	*2,026.00*

Rosbacher:	*mg/l*
Sodium	*39.90*
Potassium	*3.08*
Magnesium	*128.30*
Calcium	*255.90*
Fluoride	*0.04*
Chlorides	*48.40*
Sulphates	*7.50*
Bicarbonates	*1,442.00*

VEREINTE MINERAL UND HEILQUELLEN

This is a family company bottling **Rosbacher Mineralwasser** and **Kaiser Friedrich Quelle Heilwasser**. The first is a moderately mineralized water of exceptional purity, the other a high sodium content giving it the strong taste that has won it a loyal, but limited, following, among sportsmen and others who quench their thirst and replenish their salt level with a quick glass or two. They allied in 1983, Kaiser Friedrich Quelle having 67 per cent of the shares.

The Rosbach spring in the Taunis mountains to the east of Bonn, was known to the Romans, but Bronze Age spears and other artefacts in the museum at the source point to it as an early centre of habitation. It spouts from the ground like a geyser and is capped under glass. Bottling began in 1878, initiated by Lord Dewar of the Scottish whisky firm, who exported the Rosbach waters around the world to mingle with whisky (just as another British entrepreneur was doing with Apollinaris). Early labels directed bottles should be laid on their side (to keep the cork moist) and had testimony from Professor Wanklyn of St. George's Hospital, London that, 'The Rosbach water is remarkably pure'. The water was bought by the Appel family in 1930; they still run it. Rosbacher is sold in three versions of high, moderate and low carbonation, with the latter proving increasingly popular with the trend to *stille* water.

Kaiser Friedrich Quelle was found beneath the city of Offenbach near Frankfurt in 1888 by an industrialist looking for water for his factory. Instead he found a highly mineralized spring, useless for his machines, but wonderful to bottle. Analysis showed it had a stiff dose of 4.6 grams per litre of dissolved solids, including 1,419 mg/l of sodium. It was named after Kaiser Friedrich III, and is now sold both in the distinctive green glass of a *Heilwasser* and as a mineral water in a clear bottle.
Production: 180 million litres (47.6 million US gallons).
Bottling company:
Vereinte Mineral und Heilquellen,
Ludwigstrasse 44-62,
63067 Offenbach am Main,
Germany.

St. Michaelis: moderately and highly carbonated, low mineralization.

hella: moderately and highly carbonated, low mineralization.

ST. MICHAELIS

This relative newcomer on the German scene – bottling began only in the mid 1960s – already has many admirers. Indeed, the combined sales of **St. Michaelis** and its companion mineral water **hella** have placed the bottling company, Hansa Mineralbrunnen (thoughtfully named after the historic Hanseatic League of North Germany), in fourth place in German sales. The success is no accident; the company deliberately explored for waters that had a good, light balance of minerals, but were significantly low in sodium levels. They thus caught the trend to mineral waters packing slightly less mineral punch - and less carbonation. The aim was a water that would complement a good meal.

They found it in the small town of Trappenkamp in Schleswig-Holstein, in an unspoilt forest landscape between the North Sea and the Baltic. The area is know as the 'Switzerland of Holstein'. St. Michaelis is named after the old St. Michaelis church in Hamburg, a famous symbol of that city. The water comes from a deep channel, scoured out in the last ice age, which is covered with a 300 metre (984 feet) band of rock and sand, sealed beneath heavy, waterproof layers of marl and mica clay. The water has remarkably low mineral content by comparison with most German mineral waters, not just being low in sodium, but with only touches of calcium, magnesium and bicarbonates and a trace of fluoride. St. Michaelis is sold in highly carbonated (8 g/l) and moderately carbonated (4 g/l) versions.

The mineral content of hella is even lower, especially in sodium, so that it qualifies for use with baby formulas.

Production: over 360 million litres (95 million US gallons).

Bottling company:
Hansa Mineralbrunnen GmbH,
D-25462 Rellingen,
Germany.
Owned: Emig A.G., part of Holstein group.

St. Michaelis:	*mg/l*
Sodium	*21.00*
Calcium	*42.80*
Magnesium	*3.70*
Potassium	*3.50*
Chlorides	*10.70*
Bicarbonates	*180.00*
Sulphates	*12.10*
Fluorides	*0.21*

hella:	*mg/l*
Sodium	*8.2*
Calcium	*51.2*
Magnesium	*3.6*
Potassium	*1.3*
Chlorides	*14.0*
Sulphates	*23.0*
Bicarbonates	*146.4*

Franken Brunnen: highly or moderately carbonated, high mineralization.

St Anna: Heilwasser (health water).

St Linus: Heilwasser (health water).

Bottling company:
Franken-Mineral-und Heilbrunnen-Betriebe Hufnagel GmbH & Co. KG, Bamberger Strasse 90, 91413 Neustadt (Aisch), Germany.

Analysis:	*mg/l*
Sodium	*52.00*
Potassium	*3.52*
Magnesium	*41.60*
Calcium	*197.60*
Chloride	*72.10*
Sulphate	*338.10*
Nitrate	*0.38*
Hydrogen carbonate	*407:10*

FRANKEN BRUNNEN

Franken Brunnen sells as a staple drinking water in Franconia or middle Bavaria, and does sufficiently well in this one region to put the company into Germany's top 10 mineral water bottlers.

A bottling plant was set up in 1932 by George Friedrich Hufnagel, grandfather of one of the present directors. The factory was then in the middle of the city of Neustadt, but was moved into new plant on the edge of town in 1973. Since then two new filling plants have come into operation – one in 1989 and the other the following year – which between them can fill up to 75,000 bottles per hour.

The Franken Brunnen spring rises through limestone and is tapped in a well 70 metres (229 feet) deep in central Neustadt. The water is now piped 1.5 kilometres (1 mile) in three pipelines to the bottling plant on the edge of town. The mild-tasting Franken Brunnen mineral water is sold in greatest quantity as a highly fizzy refresher with 8.5 grams of CO_2 per litre. It is also sold as Franken Quelle with a more moderate 3.5 grams per litre of carbon gas.

The company built a factory at Bad Windsheim in 1962 to bottle St Anna *Heilwasser* (health water), and also soft drinks. In 1981, a second *Heilwasser* spring was bought, at Pechbrunn, on the Czech border, and is bottled under the St Linus label. The St Anna *Heilwasser* has a small 2 grams per litre of carbon gas from elsewhere added during bottling; and St Linus *Heilwasser* is bottled with only its own natural carbon gas, at 3 grams per litre.

Franken Brunnen has more recently acquired another three mineral waters, two in Germany and one in Hungary. In 1988 the company opened a plant in Bad Kissingen to bottle the well-known Bad Kissinger Theresienquelle water. Two years later it bought the Ileburger Schlossbrunnen factory at Eilenburg. The water has Ili Biber, a comic cartoon character associated with it. Also, in 1991, the Hungarian company, Apenta KFT, was acquired. This bottles mineral water under the Apenta label and also diet and soft drinks.
Production: more than 100 million litres (26.42 million US gallons).

Peterstaler: naturally highly carbonated, high mineralization.

PETERSTALER

This naturally sparkling mineral water from the Black Forest of south west Germany has been well known for several centuries. At home it is sold as Peterstaler, but exported, particularly to the United States, as Peters Val.

The spa at Bad Peterstal dates back to the Middle Ages and an old well, dated 1377, has been discovered in part of the sanatorium. Visitors came to drink the waters and from 1832 the water was dispatched in earthenware jugs to the surrounding towns, especially Strasbourg, just across the Rhine in what is now France. By 1866, the village of Peterstal was exporting 300,000 bottles a year.

A new bottling plant was built in 1926 by the owner of the spring, Émil Huber, and the Huber family has run the company for three generations, later taking on another source **Rippoldsauer** in the nearby village of Bad Rippoldsau.

The Peterstaler spring itself rises in abundance at a constant temperature of 15°C (59°F) in the valley of the Rench river. Carbon dioxide gas, produced in deep layers of volcanic magma, works its way up through layers of sandstone and granite fissures until it combines with a layer of water. As they surface, the water and gas work on the surrounding rocks and pick up doses of sodium, potassium, magnesium, calcium and sulphate. The resulting water has traditionally been recommended for stomach, intestinal and urinary disorders.

Peterstaler *Mineralwasser* is now marketed in the customary highly carbonated version and as a lightly carbonated *stille quelle*, while also being used as a basis for fruit drinks. Rippoldsauer, also sold as *Mineralwasser* and *stille wasser*, has slightly less mineralization. The company also bottles a *Heilwasser*, Leopoldsquelle, from another spring in Bad Rippoldsau, which is a strong mix of calcium, sodium and sulphates.

Not only is Peterstaler one of the important waters of the south west, it was also one of the first to push for more exports to the United States and the Middle East.

Production: 100 million litres (26.3 million US gallons).
Exports: a small percentage of production.

Bottling company:
Peterstaler & Rippoldsauer Mineralquellen GmbH,
Renchtalstr. 36,
77740 Bad Peterstal,
Germany.

Peterstaler:	*mg/l*
Sodium	*215*
Potassium	*17*
Magnesium	*49*
Calcium	*216*
Chlorides	*21*
Sulphates	*261*
Bicarbonates	*1,138*

Rippoldsauer:	*mg/l*
Sodium	*150*
Potassium	*10*
Magnesium	*37*
Calcium	*248*
Chlorides	*25*
Sulphates	*342*
Bicarbonates	*883*

BELGIUM

INTRODUCTION

For such a tiny country, Belgium has a very varied landscape, and in the hills and forests of the Ardennes there are mineral waters with a history of continuous appreciation since Roman times. The fame of the waters of the town of Spa has given to the world the name for a thermal station. And today the waters that once drew visitors from Russia, from England and from Sweden, among many other countries, are distributed by a large and highly automated industry.

Belgium taps no less than 38 mineral water sources and 15 spring water sources, all of which are bottled at source (Belgium adheres to the ubiquitous European 777 directive). The waters are needed. The Belgians are prodigious imbibers, whether of bottled waters, soft drinks or beer; they down more water than anyone else in Europe except the Italians, more soft drinks save the Germans, and in the beer stakes are edged out only by the Germans, Austrians and Irish. From experience, we would also add that Brussels is an excellent place to find fine wines.

On the water front, Belgium is the crossroads of Europe. Over a third of the annual production of 700 million litres (185 million US gallons) of local waters is exported.

The prime destination is The Netherlands, where such Belgian labels as Spa and Chaudfontaine are widely distributed. Indeed, they account for 80 per cent of sales there. However, the newcomer Valvert, owned by Nestlé Sources International, which was launched in 1992 with exports to France very much in mind, has already increased Belgium's role in the French market.

Imports from France and Germany, however, more than match exports. 'Acute competition' is how the local association sums it up. The French invasion has been helped by a swing towards still waters, which now account for 68 per cent of sales. While Spa is the best selling still water with almost one quarter of the market, followed by Chaudfontaine with just over 11 per cent, Contrex, Évian and Vittel from France enjoy 20 per cent of sales. Similarly in sparkling waters, although Spa and its companion carbonated water Bru are the leaders, with 27 per cent, Perrier and the Vichy waters from France, and Apollinaris from Germany are in with a nice slice, too.

Competition at home and from imports has led to concentration. Just six firms, headed by Spa and Chaudfontaine, control 80 per cent of production. The leader is Spadel SA, which markets the famous waters of Spa itself and those of Chevron and Spontin. Spa Monopole (the bottling company) is the holder of Royal Warrants as supplier of mineral water to the royal court of Belgium and the royal court of The Netherlands.

Spadel SA, under the energetic and charismatic figure of Guy du Bois, is also most active in export markets. While The Netherlands is just on the doorstep, Spadel has a considerable presence in the

Pierrot, from a 1923 poster, has become Spadel's symbol.

United States; Belgium ranks fourth in US imports, while Spa itself is the sixth best seller among imported brands. In Britain, Spadel supplies the Sainsbury supermarket chain with its own label, Monastère, and owns Brecon Carreg and Rock Spring waters. Spadel's international links also include Empresa de Aguar do Alardo in Portugal. Indeed, the big French names apart, Spadel is the most internationalized mineral water group.

National Association:
Chambre Syndicale des Eaux Minérales et de Source,
Avenue Général de Gaulle 51/B5,
B1050 Brussels, Belgium.

Source Bru: naturally moderately carbonated, low mineralization.

Chevron Monastère: still and naturally highly carbonated, low mineralization.

Sainsbury's Monastère: naturally moderately carbonated, low mineralization.

BRU-CHEVRON

Among the great trees of the Ardennes, the air alone sharpens vitality. Mile after mile of protected wood and bracken stretch way south of Liège. Deep in this forest, the bubbling waters of Chevron surface full of carbon gas, splashing the area with red stains from their high iron content.

The springs of Bru have been known since Roman times when maps were marked to show *fontanes accidii* at Chevron. From the early 7th century the spring, or *pouhon*, belonged to the Abbey of Stavelot-Malmédy, where the first bishop, St Rémacle, was credited with miracles wrought by the healing waters. The Abbey always encouraged visitors to drink the waters and, from the 17th century, supervised their export even when the nearby hotels of Spa were opposing the idea of sending water away instead of tempting visitors to come and drink on the spot. Water for export was bottled by hand in flat-sided containers, and packed in straw to be sent to Russia, France and England. The water visibly contained red splotches of iron as well as carbon gas, and some of the early bottles were brown to disguise the appearance.

In the early 18th century, the waters of Chevron outsold those of Spa, with exports of 150,000 bottles a year. After the French Revolution, the springs became the property of the commune of Chevron and quickly declined, reverting to red-stained marshlands. In the late 19th century, the commune agreed to lease the waters to a bottling company. The source in the hamlet of Bru was finally capped in 1903 and a bottling factory built directly over it. Spa Monopole bought the company in 1940.

The waters of Bru-Chevron are cool, deep waters, impregnated with carbon gas of volcanic origin. The springs surface at 10°C (50°F) forced up through shale and flint into a well dug 7.5 metres (25 feet) below ground. The springs emerge in three separate places, Source Moines, Source Sart and Source Monastère, all within a few metres of the

Testing the iron laden waters of Bru.

The source is tucked deep in the forests of the Ardennes.

bottling plant. The company now removes the iron by filtration and also the carbon gas which is stored at -25°C (-13°F) as liquid gas.

The waters are then sold in three different versions: Bru, described as *l'eau perlé* (with a few bubbles), recarbonated with 4.4 grams per litre, exactly the proportion of the original Moines and Sart springs, and bottled in plastic and glass; Chevron Monastère, a still water; or, highly fizzy, recarbonated with its own gas at 7 grams of CO_2 per litre; and Sainsbury's Monastère, which is produced for the British supermarket chain, and is recarbonated with 5.7 grams per litre, exactly the proportion of the original Monastère spring.

The label of Bru, now making fast inroads into the Belgian market as a refined table water, is emblazoned with the arms of the Prince Abbot of Stavelot-Malmédy. Director Denis Simonin says, 'Our mineral water has its own distinct personality from its history in our soil.'

Production: 36 million litres (9.5 million US gallons).
Exports: Holland, Luxemburg and the UK.

Bottling company:
SA Bru-Chevron,
Rue Bru 2,
4987 Stoumont,
Chevron,
Belgium.
Owned: Spadel.

Source Bru:	*mg/l*
Calcium	23.3
Magnesium	22.6
Sodium	10.0
Potassium	1.8
Chlorides	4.0
Nitrates	0.7
Bicarbonates	209.0
Iron	< 0.1

Ancient bottles used for exporting Spa and Bru.

Spa Reine: still, very low mineralization.

Spa Barisart: moderately carbonated, low mineralization.

Spa Marie-Henriette: naturally lightly carbonated, moderate mineralization.

SPA

Spa Reine is among the world's best-known light mineral waters, a table water harmless even to those on a salt-free diet.

Spa celebrated its 400th anniversary in 1983 as an exporter of mineral water. The waters from Spa were the first ever to travel beyond national frontiers, and the town's name has for centuries been taken as the name for a thermal station. Good water was always a matter of life and death to early communities. The naturally sparkling, iron-laden waters of Spa in a valley in the Ardennes were known to the ancient Romans, and Spa probably comes from the Latin *spagare*, to gush forth. Mediaeval legends abound of St Rémacle, the bishop of the Ardennes, and of his miracles with waters from sacred springs. By 1619 Spa boasted '400 to 500 comfortable houses for foreigners who came there from all parts of Europe to drink the medicinal waters of the place', and by 1772, 134,000 bottles of Spa water were sent abroad each year.

Spa waters are very varied, but six networks of springs produce water for bottling. The waters are shallow springs arising in three regions, Spa Centre, Spa South and Spa Nivèze. The region is part of a high plateau, or *fagne*, covered with peat and clay, but concealing below this layers of quartz shale and flint. Through these flints the waters absorb a little sulphuric acid, which attacks the carbonates and separates the carbonic acid and carbon dioxide. The acid and gas break down the rocks, and the water absorbs iron and manganese. The springs, or *pouhons*, emerge as naturally carbonated waters, containing iron and sulphur. There are more than 30 springs and boreholes in the region, of which the major ones have been capped and brought to fountains for visitors to drink. Today, the biggest seller, Spa Reine, has in fact a very low mineral content, and is a still water. Marie-Henriette, a naturally carbonated water with higher mineral content, is for export only. Barisart, with added carbon gas, sells in Belgium and for export.

Production: 476 million litres (126 million US gallons) in 1993.

Exports: to 41 countries including Holland where Spadel is a market leader.

Bottling company
Spa Monopole,
Compagnie Fermière des Eaux de Spa
34 Rue Auguste Laporte,
B-4900 Spa,
Belgium.

Spa Reine:	*mg/l*
Sodium	3.0
Potassium	0.5
Calcium	3.5
Magnesium	1.3
Chlorides	5.0
Sulphates	6.5
Nitrates	1.9
Bicarbonates	11.0
Silicates	7.0

Valvert: still, low mineralization.

VALVERT

'*Cadeau de la nature*', a gift of nature, declaimed the bright green and red packaging around six-packs of Valvert, literally 'green valley', a new mineral water that popped up on the supermarket shelves of Belgium and France in 1993. And just to drive home the message, this lightly mineralized water bubbling up from a source in the remote forests of the Ardennes was promoted as '*L'Eau à l'État Sauvage*', perhaps best translated as 'water from the wild'.

The source is indeed tucked away in a protected site of 3,000 hectares of woodland, close to the village of Étalle, just across the Belgian border from France; terrain apparently inhabited more by eagles, bears and wolves than people. The water itself percolates from north to south through light sandy soils alternating with jurassic rocks for at least 18 years before settling into the underground aquifer. The water is very lightly mineralized: a little calcium, some bicarbonates, virtually no sodium. And behind that delicate blend is the reason why Valvert suddenly appeared, to much publicity, not just in Belgium and France, but in many other countries.

Valvert represents a major effort by Nestlé Sources International to launch a new still water with low mineralization on the French market in particular, as a direct competitor to Évian and Volvic. Thus, although this is a true Belgian water, it is targeted at French drinkers; from the outset exports were set to take up 90 per cent of production. Nestlé's motive is simple; as part of the complex restructuring of the French water business when they took on the Perrier Group (see French introduction), Nestlé had to give up Volvic. This left them with Contrex and Vittel, both popular still waters, but more highly mineralized. Since the fastest growth in France is in lighter mineralization, they needed a suitable water. Valvert, which the Perrier Group had already identified in Belgium, provided the answer. 'It is a big challenge', admitted Serge Milhaud, chief executive of Nestlé's water empire. 'This is the first still water launched nationally in France for more than 20 years.'

Production: 60 million bottles, in 1.5- and 0.5-litre sizes, in 1993; target: 120 million bottles in 1994, and 180 million for 1995.

The Ardennes forests, source of Valvert.

Bottling company:
Société Générale des Grandes Sources Belges (SGGSB),
Boulevard Industriel 198,
Brussels 1070,
Belgium.
Owned: Nestlé Sources International.

Analysis:	*mg/l*
Calcium	*67.6*
Magnesium	*2.0*
Potassium	*0.7*
Sodium	*1.9*
Bicarbonates	*204.0*
Sulphates	*18.0*
Chlorates	*4.0*
Nitrates	*4.0*

SPAIN

INTRODUCTION

The last 30 years have seen an exceptional rise in the sale of mineral and spring waters in Spain, owing to the broad improvement of the economy after the Franco years, the ceaseless rise in tourism and the poor quality of tap water in many cities and on the Canary Islands. Demand increased 200 per cent in the decade 1983 to 1992.

The original market for mineral waters was in bars and restaurants, where it is natural to order *agua mineral* and wine with a meal, but the real growth has been in still waters for the family to drink at home. In the mid-1990s, still waters enjoy well over 90 per cent of the market; the traditional sparkling *agua mineral*, such as Vichy Catalán which for many visitors embodies the flavour of the Catalan region, really retains its niche in restaurants.

Meanwhile Spain's mineral water consumption is approaching an annual 50 litres per capita; with spring waters included it is closer to 60 litres. Indeed, the country is catching up fast with Europe's other great drinkers, the Italians, Germans, Belgians and French: soon after the year 2000, Spanish consumption is set to top an annual 100 litres per capita. So it is no surprise to find those big French groups Danone and Nestlé Sources International in action here with an impressive array of sources. Danone owns Font Vella, the best-selling still water.

Historically, the regulations governing the bottling of waters were copied from France, with flow controlled by the Ministry of Mines and quality control supervised by the Ministry of Health. Mineral waters had to have officially accepted therapeutic value as being *utilidad publica* (in the public interest); and they carried the designation *agua minero-medicinal*. After 1981, however, bottlers could choose to take this medicinal line, or bill the water simply as *agua mineral* with appeal to a wider market; most chose the latter. And since 1991 Spain has approved European directive 777, bringing its regulations into line with those of its European Union partners.

The Spanish association, Asociación National de Empresas de Aguas de Bebida Envasadas (ANEABE), which has 77 members marketing 96 labels, now lists under the European definition, mineral waters as 83 per cent of sales, spring waters as 13 per cent, with a small sector of treated waters. However, just 20 of those 77 companies control over 90 per cent of sales, while three groups, Vichy Catalán, and waters under the wing of Danone and Nestlé have 40 per cent.

The sparkling water market is dominated equally by Vichy Catalan on the mainland and Firgas on the Canary Islands. They enjoy 60 per cent between them, while Fonter is some way behind in third place with 9 per cent. The still water sales are much diversified. Font Vella, as league leader, has a modest 12.5 per cent, followed by the still version of Vichy Catalan at 9 per cent and

Spanish source Cardo on a rocky outcrop of the mountains near Tarragona.

Lanjaron at 6 per cent.

Clearly the impetus in Spain over the last decade has come from the French, using their expertise to put in place large distribution networks and hurrying on the trend to packaging in plastic (now nearly 80 per cent), rather than glass bottles. Consequently by 1994 Spanish sales topped 2 billion litres (528 million US gallons) for mineral waters and over 300 million litres (79 million US gallons) in spring waters.

The prime producing region is Catalonia. Well over half of all waters bottled in Spain come from the hill villages between Barcelona and the Pyrenees. Here, too, are sources of volcanic origin which the Romans first exploited as thermal baths in villages such as Caldes de Malavella, where the first more modern spas flourished 100 years ago.

The Canary Islands also have volcanic sources producing water such as Firgas which provide a pleasant alternative to poor municipal supplies.

And, just as in Spain you will usually encounter only domestic wines, so it is with mineral waters. Imports are almost negligible, but then so are exports. This is a country in which to enjoy the local waters.

National Association:
Asociación Nacional de
Empresas de Aguas de
Bebida Envasadas (ANEABE),
C/Velazquez 54, 50C,
28001 Madrid,
Spain.

Font Vella: still, low mineralization.

Fonter: naturally carbonated, low mineralization.

FONT VELLA

Font Vella, in the Catalán dialect of north east Spain, means 'beautiful spring'. And this clear, refreshing water, the top seller in Spain, indeed comes from a picturesque bowl in the wooded hills of Montseny-Guilleries in the province of Gerona.

Actually getting to the little town of Sant Hilari, which has been known as the town of 100 sources since the 19th century, underlines how remote it is from modern life. A mountain road (negotiated with astonishing dexterity by the drivers of Font Vella's delivery trucks) twists up through chestnut groves and past mountain streams to emerge finally at an altitude of 800 metres (2,625 feet) in a town of 4,000 people that prides itself on wood carving and waters.

The Font Vella bottling plant, one of the most modern in Spain, sits atop a huge aquifer about 100 metres (328 feet) down in the granite hill. The full extent of the aquifer is still being explored, for there are many side galleries in a very fractured rock structure. But the sheer hardness of the granite explains why the water is so well filtered and so lightly mineralized (it has only modest amounts of silica, bicarbonates and calcium). Originally the water came only from a small spring where villagers still gather to collect it in jugs and bottles, but the bottling company has now bored down three capped pipelines, through which the water is pumped into four huge stainless steel tanks, each holding 0.5 million litres (0.13 million US gallons), set into a gallery carved out of the rock beneath the factory. They have also put down two experimental bores in an attempt to determine the limits of the aquifer. But the flow is held at 1.5 million litres (0.4 million US gallons) a day maximum to avoid reducing its water level.

Font Vella was originally approved *utilidad pública* in 1956. Its real success, however, has come since the 1970s, when BSN (now Danone) the French food and drinks company, which also owns Évian, bought a controlling share in EBAMSA (Explotación de Balnearios y Aguas Minerales SA) which owns Font Vella and Fonter, a naturally carbonated water from a neighbouring hill town. The French group at once put in management with experience from Évian and Badoit, so that in terms of technology and quality control EBAMSA was effectively unrivalled in Spain. These credentials made Font Vella well placed to benefit from the rapid growth of still sales. Font Vella now has 12.4 per cent of still sales, comfortably ahead of all rivals. Although its appeal is broad, it has also cut an image as a slimmer's mineral water, with advertisements showing a tape measure around a wisp of a waist and the slogan *'Si le preocupa la silueta – el agua ligera'* (if you are preoccupied with your figure – the light water).

Its companion water, **Fonter**, from the village of Gerona, has a modest 9 per cent share of sparkling sales. The source here has been known for centuries as Font Picant, because of its natural effervescence. The waters bubble out of a fault in the valley of the river Brugent, just outside the village. Although this is a region of numerous geological faults with many springs of distinctive character, only Font Picant emerges with natural carbonation. The water is pumped through stainless steel pipes to the bottling plant just 50 metres from the source. Fonter is lightly mineralized with bicarbonate and calcium, but has a very low sodium content. Although its bottling was first officially authorized in 1903, it was finally approved as an *agua minero-medicinal* in 1957. It is bottled with a relatively high level of carbonation at 7.5 grams per litre and distributed nationally. It is good for helping digestion, especially after some of those heavy Spanish meals.

Production: Font Vella 250 million litres (66 million US gallons); Fonter 15 million litres (4 million US gallons).

Bottling company:
EBAMSA,
Apartado 7004,
Gran Via 324,
08011 Barcelona 4,
Spain.
Owned: Danone (BSN).

Font Vella:	*mg/l*
Bicarbonates	109.8
Sulphates	7.5
Chlorides	5.9
Calcium	25.6
Magnesium	4.9
Sodium	11.8

Fonter:	*mg/l*
Bicarbonates	130.3
Sulphates	17.8
Chlorides	9.0
Calcium	35.3
Magnesium	7.3
Sodium	10.9

Snowbound source high in the hills of Gerona.

Vichy Catalán: naturally carbonated, high mineralization.

AGUAS MINERALES ACÍDULAS Y BICARBONATADAS ALCALINAS

DEL MANANTIAL

VICHY CATALAN

DECLARADAS DE UTILIDAD PÚBLICA

por R. O. de 5 de marzo de 1883

CALDAS DE MALAVELLA (GERONA)

ESPAÑA

TEMPERATURA: 60 GRADOS EN LA EMERGENCIA

Gasificada. Contenido en anhídrido carbónico: 5,5 gramos por litro.

PREMIOS OBTENIDOS

Medalla de Plata	Nice	1884
Medalla de Oro	Nápoles	1884
Medalla de Plata	Zaragoza	1885
Medalla de Plata	Zaragoza	1886
Medalla de Oro	Barcelona	1888
Medalla honorífica	Paris	1889
Diploma	Badalona	1892
Gran Premio	Barcelona	1929

APLICACIONES

Las indicaciones de las aguas de *Vichy Catalán* son las propias de las aguas bicarbonatadas sódicas carbogaseosas. Especialmente las gastropatías hipersecretoras, gastroenteropatías y hepatopatías en que resulte favorable la acción alcalina de estas aguas. Asimismo son útiles como tratamiento coadyuvante de la diabetes y como favorecedoras de la eliminación del ácido úrico. Contienen flúor.

VICHY CATALÁN

Vichy Catalán is Spain's best-selling naturally carbonated mineral water controlling 30 per cent of the sparkling market. Its rich blend of bicarbonates, chloride and sodium gives it a distinctive flavour, somewhat akin to Vichy Celestins in France after which this Catalán water was originally named a century ago.

You can still sit on the steps of the Roman baths at Caldes de Malavella just south of Gerona in Catalonia and imagine the Romans sipping the effervescent, deep volcanic waters that bubble up all over this little town at a steamy 60°C (140°F). Although all are from the same aquifer, their composition does vary slightly. The waters are now bottled under four labels: Vichy Catalán, Imperial, San Narciso and Malavella. But Vichy Catalán is the best known. After the Romans, the merits of this water were first recognized by a Barcelona doctor in the 1880s, who built a small spa at Caldes de Malavella in 1883, and named it Vichy Catalán after the French thermal resort. Bottling started in a small way around 1900, and the water was distributed through pharmacies. The spa enjoyed its social peak during the 1920s, but was then badly damaged in the Spanish Civil War, when it served as a hospital. More recently, it has been completely modernized with all the facilities for thermal baths, water massage and hydrotherapy and is seeking to attract a new clientele of health- and diet-conscious Spaniards.

Meanwhile a modern bottling plant has boosted production to permit national distribution. The single source has been capped, and the water and natural carbon gas are separated as they emerge, and stored initially in tanks while the water cools. The gas is then re-injected on the bottling line at 5.5 grams per litre. In recent years a still version of the water has also been launched, and it has been selling well.

Production: 180 million litres (47 million US gallons).

Bottling company:
Vichy Catalán,
Lauria 126,
08011 Barcelona,
Spain.

Analysis:	*mg/l*
Bicarbonates	*2,104.5*
Chloride	*620.0*
Sulphide	*48.9*
Fluoride	*7.3*
Sodium	*1,133.3*
Potassium	*48.0*
Calcium	*32.9*
Magnesium	*7.8*

IMPERIAL
CALDES DE MALAVELLA
0,25 LITROS

San Narciso: naturally carbonated, high mineralization.

Imperial: naturally carbonated, high mineralization.

SAN NARCISO/ IMPERIAL

Just down the road from Vichy Catalán and right next to the old Roman baths, is the bottling plant for San Narciso and Imperial. Both these sources are owned by Eycam Perrier SA, part of Nestlé Sources International. Together they have a modest 4 per cent of sparkling sales in Spain.

Although both waters are almost identical to Vichy Catalán, Imperial has a slightly higher phosphate content, while San Narciso has a tinge of sulphur in its flavour. And the local villagers apparently prefer San Narciso; at least the old ladies may be seen in the mornings taking their glass of hot water as it trickles from a little fountain in the street. Both are bottled with a slightly lower gas content, 4.5 grams per litre, than Vichy Catalán. San Narciso has been bottled since 1870, and won its *utilidad pública* approval in 1928, and Imperial was officially accredited in 1916.

But the flow of both sources is rather limited: San Narciso at 3,000 litres (792 US gallons) an hour, and Imperial a mere 900 litres (238 US gallons). So Imperial has limited distribution only in Catalonia, although San Narciso may be encountered more widely in the main cities of Spain.

Production: San Narciso and Imperial: 12 million litres (3.2 million US gallons).

Bottling company:
Explotación y Comercialización de Aguas Minerales SA (Eycam),
Aragon 212,
08011 Barcelona,
Spain.
Owned: Nestlé Sources International.

San Narciso:	*mg/l*
Bicarbonates	2,147.0
Sulphates	606.4
Fluoride	8.0
Sodium	1,120.0
Potassium	50.0
Calcium	52.9
Magnesium	8.7

AGUA MINERAL NATURAL

Viladrau

Manantial Fontalegre

MACIZO DEL MONTSENY

Análisis químico (en mg/l.):
Residuo seco 113,7 -HCO₃ 68,3 - SO₄ 8,0
Cl 8,4 - Ca 20,0 - Mg 2,7 - Na 8,2 - SiO₂ 21,4
Lab. Dr. Oliver Rodés/Barcelona, mayo 1991

Agua de baja mineralización. Mineromedicinal.
Declarada de Utilidad Pública 4/12/1973
Registro Sanitario 27.19/GE 23
Manantial en Viladrau (Girona)

INDICADA PARA DIETAS POBRES EN SODIO
Controlada sistemáticamente por el laboratorio de análisis instalado en la propia Planta Embotelladora.
Proteger de la luz solar. Conservar en lugar fresco, seco y apartado de olores intensos.

CONSUMIR PREFERENTEMENTE ANTES DE FIN DE 1998

EYCAM-PERRIER, S.A. - Aragón, 212 - 08011 BARCELONA

Agua de Viladrau: still, very low mineralization.

AGUA DE VILADRAU

This water from the source La Cuaranya (meaning where two mountains meet) high in the Serra del Montseny of Catalonia is one of the least mineralized in all Spain. It is also a relative newcomer, having only been declared *utilidad pública* in 1973. Apparently this spring, nearly 1,000 metres (3,280 feet) up in the mountains, was traditionally used to water nearby fields, until a family of local farmers named Serrat started bottling the water on the advice of a son who qualified as a chemist.

But the real development came when Perrier bought a stake in 1979. A tunnel 42 metres (137 feet) long has now been driven into the hillside, and the water is collected at three points along it, and channelled by gravity to the new, modern bottling plant 50 metres (164 feet) below.

The flow varies considerably, and this was one of the very few sources we visited where the rate could be quickly related to recent rains. After a very heavy storm, the flow will start to increase within two days, and rise for two weeks, after which it stabilizes and slowly subsides. However, this does not mean the rainwater itself percolates so quickly, rather that general pressure is applied on the surrounding water table.

The low mineralization is accounted for by the very old fractured granite segments within the hill which are now almost like sand, thus acting as a big natural filter.

Eycam-Perrier has substantially increased production over the last decade. When we first toured the plant in 1984 its capacity was 30 million litres; today it is nearly triple that. Viladrau has caught the trend for light waters in Spain.

Production: 80 million litres (21 million US gallons)

Bottling company:

Eycam-Perrier SA,
C/Aragon 212,
08011 Barcelona,
Spain.
Owned: Nestlé Sources International.

Analysis:	*mg/l*
Bicarbonates	58.6
Chloride	6.0
Sulphates	6.7
Nitrates	7.5
Sodium	8.8
Potassium	1.0
Calcium	15.6
Magnesium	2.2

LANJARÓN

The waters of Lanjarón, which rank third in still sales in Spain, come from the region of Alpujarra high in the Sierra Nevada between Granada and the Costa del Sol. Originally one of the first spas in Spain was established here in 1869, taking advantage of the pleasantly temperate climate halfway between the mountain peaks and the sea. To this day it remains one of the most frequented spas in Spain, with 5,000 *curistas* coming to take the waters every summer. There are several sources around the village, some of them highly mineralized, and their waters are taken only under medical supervision. Only two with moderate mineralization are actually bottled. Bottling first began in the 1930s so that water could be sent to *curistas* on an informal basis, but did not start on a commercial basis until the 1960s. Lanjarón has national distribution (although it will be seen most widely in tourist centres of the Costa del Sol). The company has pioneered the PET plastic bottles in Spain, in which most of its water is shipped. Lanjarón once had a good share of the restaurant and bar trade, but now sees much greater potential in developing sales for home consumption, because of the poor quality of tap water in many Spanish cities. A small part of the production is carbonated.
Production: 100 million litres (26 million US gallons).

Lanjarón: still and carbonated, medium mineralization.

Bottling company:
Lanjarón, Apartado 777, Granada, Spain.
Owned: Danone (BSN).

Analysis:	*mg/l*
Bicarbonates	*160.0*
Sulphates	*29.0*
Calcium	*50.0*
Magnesium	*12.0*
Potassium	*1.0*

CARDÓ

Agua del Valle de Cardó comes from a spectacular rocky outcrop in a semi-circle of mountains south-east of Tarragona. A Carmelite abbey on the rock spur became famous for its healing waters in the 18th century, and a spa thrived there from 1880 until 1965. Although the waters were declared *utilidad pública* in 1887, commercial bottling began only in 1973. Rainwater from the surrounding hills filters down through layers of dolomite and limestone to form a large aquifer above a bed of clay from which it emerges naturally at two springs, Borboll and Pastor. They have a combined flow of 24,000 litres (6,340 US gallons) an hour of very lightly mineralized water.
Production: 12 million litres (3.2 million US gallons).

Cardó: still, low mineralization.

FIRGAS

The leading mineral water of the Canary Islands comes from the village of Firgas in the hills 25 kilometres (15 miles) north of Las Palmas on Grand Canary. It is of volcanic origin and emerges from one natural source, La Ideal, at a warm 21°C (69.8°F) with light natural effervescence (1.5 grams per litre). Originally declared *utilidad pública* in 1929, production has since been increased by drilling two *sondages* down to 230 metres (755 feet). Although lightly mineralized, it does have a relatively high silica content. Most sales of La Ideal are naturally carbonated, an artificially de-carbonated version, La Ideal 1, is also sold.
Production: 50 million litres (13 million US gallons).

Firgas: naturally carbonated, low mineralization.

PORTUGAL
INTRODUCTION

'Ours is a country of mineral waters,' remarked Dr B M d'Almeida, former director of Pedras Salgadas, one of the best-known Portuguese waters. Indeed, Portugal regards its mineral waters as one of its national treasures. Many of them have been known for their therapeutic qualities since Roman times.

'For 2,000 years people have journeyed every summer to the source at Chaves, which the Romans called Aqua Flavia, to take the water,' added Dr d'Almeida proudly. In the heyday of the Portuguese empire in the 15th and 16th centuries, the king and his entourage went in the summer to Caldas de Monchique in the Algarve to restore their health and cure their *melancolico* by sampling Monchique's waters. The *Aguas Santas* (Holy Waters) of Vimeiro, on the coast near Lisbon, were equally renowned for their medicinal qualities. The king's physician wrote in 1726 that the waters from Vimeiro's fountain 'provide an excellent cure for nephritic pains, since they possess a gift of disintegrating and expelling the stones and sands that have become lodged in the kidneys or spleen'. By the late 19th century, Portugal boasted some of the finest spas in Europe, with grand hotels, casinos, bathing establishments and drinking halls. Probably none was finer than the ornate drinking hall at Melgaço, in northern Portugal, illustrated here, with its characteristic wrought iron lattice-work portals and windows.

That tradition has lasted. Even today there are no less than 44 spas in Portugal and the islands of the Azores where over 150,000 people spend a week or two in the summer taking their cure. Many of the sources are of volcanic origin and surface along a 300-kilometre (186-mile) fracture running from north São Petro do Sul through the beautiful and remote Sabroso de Aguiar valley across the border into north west Spain. The waters emerge where secondary fractures cut across the main fissure. Although most are relatively highly mineralized, the character of each, even from sources only a few kilometres apart, is distinctive.

The medical message, incidentally, was carried through to labels once the waters were bottled, for many were described as *agua minero-medicinal*. That title remains today with the trade association, Associação Nacional dos Industriais de Águas Minero-Medicinais e de Mesa, to which all bottlers and most spas belong.

The ornate drinking hall at Melgaço in Northern Portugal.

Old Melgaço bottle label.

Surprisingly, perhaps, relatively few spas bottle their waters. Although 16 mineral water companies, plus another 10 marketing spring waters, distribute bottled waters throughout mainland Portugal and the Azores, in practice four of them dominate the market. Luso, the leading still label, alone accounts for over 35 per cent of all sales. Overall still waters account for over 80 per cent, reflecting a steady shift as the Portuguese gradually look to bottled water more as a thirst quencher than as a health cure. Some traditional naturally carbonated favourites, however, remain, notably Pedras Salgadas and its associated sources, Vidago and Melgaço.

Although sales have increased in recent years, rising to 286 million litres (75 million US gallons) of mineral water by 1992, the per capita consumption remains low by comparison with most of Europe at just under 30 litres, a third less than neighbouring Spain. This is a legacy of Portugal's late economic growth. The gap, however, is narrowing, and is actually less if bottled spring waters are included; they topped the 100 million litre mark by 1992 (a seven fold rise since we first wrote this guide) and now claim over 25 per cent of the water market. The potential for further growth, the industry argues, is considerable, at least to match Spain's drinking habits over the border. That potential has already attracted foreign investors: Nestlé Sources International owns two waters – Castello and Campilho, and Belgium's Spadel has Alardo.

Despite a determined push in the mid-1980s to promote exports, notably by Pedras Salgadas to get niches in the United States and Britain, they remain insignificant at less than 3 per cent of output.

The original Portuguese regulations on mineral waters were drafted in 1929, but since 1992 the full European directive 777 has been adopted to bring the country into line with its European Union partners. In fact, the rules were already tight, requiring official Ministry of Health approval of their therapeutic benefits and bottling under conditions of strict hygiene at the source. In an age of increasing pollution, one of the real strengths of the Portuguese waters is that their sources are virtually all in remote rural areas.

National Association:
Associação Nacional dos Industriais de Aguas Minero-Medicinais de de Mesa, Avenida Miguel Bombarda, 110 2 DT, 1000 Lisbon, Portugal

Pedras Salgadas: naturally lightly carbonated, high mineralization.

PEDRAS SALGADAS

Pedras Salgadas is a strongly bicarbonated, alkaline mineral water that is renowned in Portugal as an aid to digestion. Its relatively low level of natural carbonation (only 2.5 grams per litre) makes it comparable to Badoit in France or Bru in Belgium as a natural accompaniment to good food and wine.

The village of Pedras Salgadas lies in the Sabroso de Aguiar valley of northern Portugal beneath the moorlands of the Alvão plateau, from which rain and snow filters down. Pedras, like its neighbouring source of Salus Vidago 12 kilometres (7.5 miles) up the valley, and the ancient spa of Chaves just beyond, is astride the great volcanic fault running for 300 kilometers from the heart of Portugal up to the Spanish border. All the waters of this valley are of deep volcanic origin, relatively high in mineral content and naturally carbonated.

Pedras Salgadas first enjoyed fame as a spa in the 1880s, even attracting the Portuguese royal family in the summer time. Actual bottling started early in this century. The water emerges as six springs christened Don Fernando, Gruta Maria Pia, Penedo, Grande Alcalina, Preciosa and Pedras Salgadas. But several are rather saline, and only Pedras Salgadas, with the highest blend of minerals, is bottled. Pedras now comes from five capped boreholes drilled down to between 80 and 110 meters (262 and 361 feet) from which there is a constant flow of 150,000 litres (39,630 US gallons) per day at 16°C (60.8°F). All the water is brought up initially into glass reservoirs, here the water and gas are temporarily separated into storage tanks.

For three generations Pedras Salgadas was owned by the local d'Almeida, d'Oliveira and Serodio families, but they sold out in 1982 to José Sousa Cintra, a Portuguese property developer. By then, too, an alliance had been formed, not only with Salus Vidago nearby but also with Melgaço and Lombadas, the main mineralwater in the Azores. Thus the combined group, Vidago, Melgaço, Pedras Salgadas SA, now controls virtually all the naturally carbonated waters in Portugal.

At Vidago there are four springs, Vidago I, Vidago II, Oura and Salus. Both Vidago sources are used exclusively for therapeutic purposes, with very limited bottling, while the waters of Salus and

Natural gas surges up in the spring at Pedras Salgadas.

Oura, with a flow of 50,000 litres (13,208 US gallons) daily, are bottled for general distribution under a Salus Vidago label. The water has a slightly higher sodium bicarbonate content than its neighbour Pedras Salgadas and has slightly less natural carbonation. Melgaço, close to Portugal's northern border with Spain, began as a spa in the 1880s, offering visitors a more moderate and rather wider 'basket of minerals' than Pedras Salgadas or Vidago; the naturally carbonated water from its twin fountains, Fonte Principal and Fonte Nova contains much less sodium bicarbonate, but blends in more calcium and magnesium. Originally it was recommended for diabetics. Bottling today is limited, and it is not widely seen in Portugal.

Lombadas is a naturally carbonated, lightly mineralized water from a remote valley of San Miguel island in the Azores. It has a distinctive acidic flavour.

Production: Pedras Salgadas: 26 million litres (6.7 million US gallons).
Salus Vidago: 10 million litres (2.6 million US gallons).
Melgaço: 0.6 million litres (0.16 million US gallons).
Lombadas: 0.5 million litres (0.13 million US gallons).

Bottling company:
Vidago, Melgaço & Pedras Salgadas SA, Auto Estrada Lisboa-Sintra, Km 2 Alfragide, 2700 Amadora, Portugal.

Pedras Salgadas: *mg/l*	
Chloride	*40.50*
Sulphate	*2.30*
Bicarbonates	*1,866.60*
Fluoride	*2.22*
Sodium	*549.70*
Potassium	*25.80*
Calcium	*132.00*
Magnesium	*8.50*

Salus Vidago:	*mg/l*
Chloride	*37.90*
Sulphate	*2.60*
Bicarbonates	*2,027.40*
Fluoride	*4.25*
Sodium	*660.10*
Potassium	*44.60*
Calcium	*78.00*
Magnesium	*10.30*
Iron	*0.06*

1930's painting shows Pedras Salgadas as a fashionable spa.

REGIONAL WATERS

LUSO

Luso is Portugal's best-selling still mineral water and is drunk essentially as a very light table water. It comes from a source mid-way between Lisbon and Porto which first flourished as spa in the 1850s, and even today is a substantial resort with extensive facilities for hydrotherapy.

The water, with a very low mineral content compared to most others in Portugal, comes from an aquifer 70 metres (230 feet) down in quartzite rock with a regular flow of 42,000 litres (11,095 US gallons) hourly. Termas do Luso is 53 per cent owned by Central de Cervejas SA, with the balance held by a foundation set up by Dr Byssaia Barreto for the benefit of homes for children and the elderly. Luso enjoys virtually 37 per cent of the entire water market in Portugal. 'Our water is drunk, not for medicinal reasons, not even for thirst, but as the favourite table water for Portuguese families,' the director told us.
Production: over 105 million litres (27.7 million US gallons).

ALARDO

The Alardo spring rises on the south-eastern slopes of the Serra da Gardunha, a remote region in central Portugal, not far from the Spanish border. The water gushes up through a deep fracture in the granite rock at a rate of 20,000 litres (5,300 US gallons) an hour, and surfaces at a cool 13°C (55.4°F). It is carried by force of gravity from the spring to the nearby bottling plant. A soft, still water, Alardo contains little sodium and few other minerals. But during its underground journey through ancient rock strata it acquires minute quantities of an exceptional number of rare trace elements, many of which are valued for their influence on human cellular development.

The Alardo enterprise was taken over by Spadel, the Belgian conglomerate, in 1990, and since then production has increased by almost 50 per cent. Alardo is now the second largest selling brand of mineral water in Portugal.
Production: 43 million litres (11.4 million US gallons).

PIZÕES/CASTELLO

The water from Pisões-Moura, in south eastern Portugal, has a moderate mineral content of bicarbonate and calcium. It is sold in two varieties under two different brand names. Pizões, the still version, is bottled at source in its natural state, and sold in three sizes of PVC bottle. Highly carbonated Castello has the fizz added during the bottling process and is available only in glass. Vittel, which is now owned by Nestlé Sources International, has had a stake in the Portuguese company since 1981.
Production: Pizões 5 million bottles; Castello 30 million bottles.

CARVELHELHOS

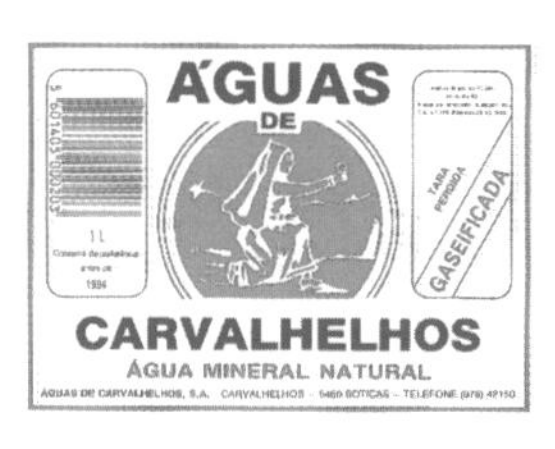

Carvelhelhos is a bicarbonated mineral water from the village of Boticas in the Serra de Alturas mountains just above the valley where Pedras Salgadas and Salus Vidago emerge. But, unlike its neighbours, the water is not naturally carbonated and is lightly mineralized. Since its discovery in 1850, it has been a small spa, billing its waters, rather charmingly, as 'attempting to cure' skin maladies, along with the usual digestive benefits. But it is an acceptable table water with a low salt content and the added benefit of fluoride.
Production: 22 million litres (5.8 million US gallons).

FASTIO

One of Portugal's lightest mineral waters, Fastio is a comparative newcomer to the market, having been bottled only since 1979. This still water comes from a cave in the granite mountains of Serro do Geres in northern Portugal, and the company claims to be the first to have used 1.5-litre (0.4 US gallon) plastic bottles. Although mineralization is low, Fastio's silica content is relatively high. It is now the fourth most popular brand in Portugal, and is also exported, mainly to the US.
Production: over 15 million litres (4 million US gallons).

CAMPILHO

Campilho is a moderately mineralized, naturally carbonated water from a source at Vidago in the mountainous Alto Douro region of northern Portugal. Its beneficial qualities came to be recognized throughout the region, and commercial exploitation of the spring began more than a century ago. In 1988, Vittel, now part of Nestlé Sources International, acquired an interest in the bottling company. Campilho is distributed mainly in the north of Portugal.
Production: 9 million glass bottles.

MONCHIQUE

The waters of Monchique from four springs in a faulted mountain valley near Faro in the Algarve in southern Portugal were known to the Romans, and the Portuguese royal family went there for health cures as early as 1495. Originally the sources were owned by the church, which gave them to the state in the mid-19th century. The spa, which still treats nearly 2,000 people annually, is now operated by the National Tourist Board. The water from two of the sources with a relatively high sodium content has been bottled since 1959, and was initially favoured in contrast to the poor quality of local water.
Production: 8.8 million litres (2.32 US gallons).

SWITZERLAND
INTRODUCTION

As the centre of some of the richest food in Europe, and renowned for the Alps, Switzerland has long offset digestive stresses with draughts of pure mountain water. Although the spa tradition has declined, most Swiss families today have bottled water on the table at home, and restaurants bring mineral water almost automatically with wine.

Swiss drinking habits, however, do show a fascinating contrast compared to some of their European neighbours with an exceptionally balanced approach. Thus the Swiss annual per capita consumption of bottled waters is 76 litres, of soft drinks 77 litres, and of beer 71 litres; compare that to the Germans, who drink 147 litres of beer to 93 litres of bottled waters, or Italy where water consumption is 116 litres, beer a mere 40 litres and soft drinks 22 litres. This is a revealing insight of national tastes. Not that the Swiss are laggards on water; mineral water production nearly doubled between 1983 and 1992 to reach 455 million litres (118 million US gallons) and they rank fifth equal with the Austrians in per capita consumption, only a few glasses behind the French.

Their taste, though is more Germanic; over 85 per cent of the waters bottled in Switzerland are carbonated, although a trend in favour of lightly carbonated or still waters has grown in the 1990s. Only in the French-speaking cantons are still waters really preferred, and then the big French waters like Evian (which is just over the French border from Geneva), Vittel or Contrex dominate. Aside from those imports, and San Pellegrino from Italy, which in all account for around 14 per cent of the market, the Swiss bottled water business is a domestic business; exports are a mere 2.5 per cent of production.

Although there are 25 recognized mineral water sources controlled by 18 companies – including several breweries – only 12 waters have substantial distribution. If you are travelling in Switzerland, you are most likely to encounter Henniez (or its new still water Cristalp), Passuger, or Valser, although many other own brand labels now show up on supermarket shelves, often at half the price of the famous names.

Although Swiss regulations on both food and mineral water have been largely brought into line with the European Union (even though Switzerland does not belong), and hygiene controls are tough, bottlers do have some leeway compared to, say, the French. No Swiss spring, for instance, surfaces with more than 1 gram of carbon gas per litre, but many sell with a stiff 5 or 6 grams. The extra gas is brought from such places as the Black Forest in Germany to be injected on the bottling line. However, the label must state 'added CO_2'. The Swiss bottlers are permitted to treat their waters to remove sand, iron, manganese and sulphur, while an overdose of fluoride can be reduced (over 50 per cent of Swiss waters need such treatment). They may also use mineral water sources to make soft

The Vals valley, home of one leading mineral water.

drinks; an important factor for a group like Henniez which markets Henniez Orange and other fruit drinks using water from any of its five springs. However, Swiss law does thwart the French and German tradition of therapeutic waters, because any water with medicinal claims must be registered as a drug and may not advertise. So fully fledged medicinal waters have a tiny share of the market. But a leading Swiss water like Valser can market itself as a *Heilwasser* in Germany.

The challenge facing Swiss bottlers over recent years has been the widespread sale of own label waters at prices which scarcely cover costs. The original success came with the Aproz label of the Migros supermarket chain. Other supermarkets then sought out recognized mineral water sources, and signed them up for own label. So the famous names suddenly found their bottles sharing shelf space alongside their own water under a supermarket ticket and selling at half price, a fact not unnoticed by customers spying a bargain. But the Swiss, as in banking, are rather secretive on the market success of own label. 'It has not yet been found possible to determine reliably the market share of these low price products,' observes an industry report.

Enterprise and a good water, however, can overcome such obstacles. The story of Aqui, the mineral water discovered in the middle of Zurich by the Hurlimann Brewery when it was drilling for extra sources for its beer production, shows that a new water can still break into the market. And the success of Valser, whose mildly carbonated mountain water arrives from friendly door-to-door delivery vans throughout the German-speaking cantons, reaching top place in home consumption, have proved it is possible to get your water to the customer in defiance of middlemen and conventional outlets.

National Association:
Verband Schweizerischer Mineralquellen,
General Wille Strasse 21,
Postfach 307,
8027 Zurich,
Switzerland.

Henniez: still and moderately or highly carbonated, moderate mineralization.

Cristalp: still, moderate mineralization.

HENNIEZ

Henniez is the best-selling mineral water in Switzerland, holding almost a third of the entire market with its still, lightly carbonated and carbonated labels. Henniez emerges from the hillside in the valley of the Broye near Lausanne, and is filtered down through sand and gravel from the foothills of the Alps to an aquifer between glacial moraines.

Henniez prides itself on Roman origins, and Napoleon is also said to have sampled the renowned waters when his armies passed by on his campaigns. A guide to the Vaud region in 1867 observed, 'Above the village of Henniez is a romantic site where you find baths frequented for more than five centuries by the inhabitants of the district.' That local reputation was capitalized on in 1880 by a Dr V. Borel from nearby Neuchatel, who decided to exploit it as a spa. L'Hotel Des Bains was built and Swiss high society came to be treated for liver and kidney ailments and rheumatism with the alkaline water that contains bicarbonate, calcium and magnesium. A few years later, in 1905, a Société Bains et Eaux d'Henniez Lithinée SA was set up, and the first bottling plant established.

To begin with, two sources in the hillside were tapped: Bonne Fontaine and Espérance. A third, Alcalina, was opened in 1927, and since then two others have been added. All five are within a square kilometre in a valley shaded by beech and pine trees. At each source tunnels have been drilled back into the hillside to tap the natural flow of the aquifer. The water, which emerges at a cool 9.2°C (48.5°F), passes down stainless steel pipes to the new bottling plant 2 kilometres away. The composition of waters from each source is virtually identical, so that they can be blended.

Although Henniez is a still mountain water, an artificially carbonated version containing 7 grams per litre of carbon gas (permitted under Swiss regulations) has long been produced. There is also a lightly carbonated version, Henniez Legère, containing only 4 grams of CO_2 per litre.

Cristalp. In 1989 the Henniez group acquired another source of natural mineral water at Saxon, a former spa town located in the centre of the

Valaisan Alps, which during its heyday in the late 19th century attracted celebrities such as Dostoyevsky and Garibaldi to its thermal centre.

The Saxon source, believed to be one of the most bountiful in Europe, emerges at a constant temperature of 25°C (77°F) from deep in the alpine rock, through which it has taken 25 years to filter, at a rate of 2,500 litres (645 US gallons) per minute.

The Source by the Crosses, as it became known from the crosses planted nearby by grateful visitors whose skin disorders, digestive complaints and other ailments had been alleviated by bathing in or drinking the water, was rediscovered in 1971 after a period of decline, and commercial activity was resumed on a small scale in 1983. When Henniez took over the bottling plant at the end of the 1980s, it launched its new brand on the Swiss market as Cristalp, which is exported under the name of Alpwater. Cristalp is a moderately mineralized, exceptionally pure non-carbonated water that is rich in fluoride and contains a useful dash of iodine, helpful in preventing goitre. Cristalp projects a healthy, environmently friendly image which has made it extremely successful in its country of origin. Since it first appeared in 1989, Cristalp has captured 13.5 per cent of the Swiss market for still water. More than 11 million bottles of Cristalp were produced in 1992. The company has invested 20 million francs in enlarging, modernizing and automating the bottling plant.

Production: Henniez about 142 million litres (36.6 million US gallons). Cristalp 12.5 million litres (3.2 million US gallons).

Exports: to Belgium, The Netherlands, the United Kingdom, the United States.

Inside the Source Alkalina at Henniez.

Bottling company:
Source Minérales
Henniez SA,
1525 Henniez,
Vaud, Switzerland.

Henniez:	*mg/l*
Calcium	111.20
Magnesium	19.20
Sodium	8.50
Bicarbonates	374.50
Nitrates	21.60
Chlorides	16.80
Sulphates	13.80
Fluorides	0.07

Cristalp:	*mg/l*
Calcium	115.00
Magnesium	40.00
Sodium	19.90
Potassium	1.80
Bicarbonates	306.00
Chlorides	11.50
Nitrates	1.80
Sulphates	211.00
Fluorides	1.40

Valser St Petersquelle: still and moderately carbonated, high mineralization.

VALSER ST PETERSQUELLE

Rich in minerals from the remote Vals valley in the Alps, and named after the local saint, Valser St Petersquelle is a powerful draught which has quickly established itself as a leader among Swiss bottled waters.

The spring near Chur was renowned as early as 1732, but it was then used primarily as a thermal bath. During the 19th entury a hotel and thermal spa were still in operation. However, the hotel eventually closed and the spring was put up for sale. Then in 1960 the water was re-discovered by an enterprising team of two Swiss and a German. They found that the water, surging up at 30.2°C (86°F) from an artesian well 1,000 metres (3,200 feet) deep at a plentiful 500 litres (132 US gallons) per minute, was mineralized at 1,800 mgs per litre. Tritium tests also showed it had spent at least 20 years in the mountains.

The partners have now built a bottling plant in the valley below the source, and piped the water down to a reservoir there. In the factory iron and manganese are removed from the water by filtration, and natural carbon gas brought from the Black Forest is injected at the rate of 4 grams per litre. A still version was marketed in 1983 to appeal to the French-speaking cantons where nearly half the market is for still water. The overall flavour of the water is of its high calcium and magnesium content.

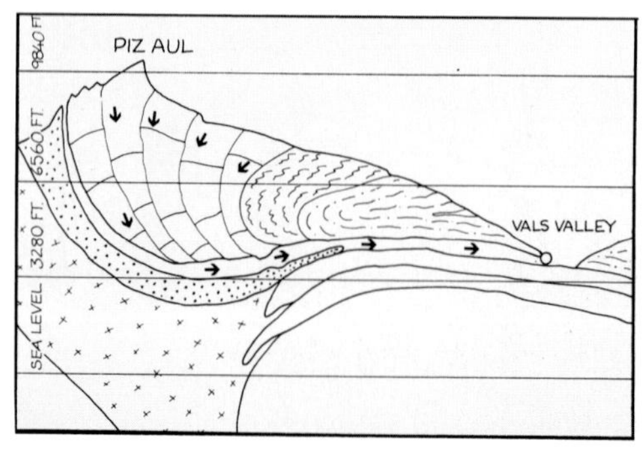

Breaking into the Swiss market with a new water in the 1960s posed the owners an interesting challenge. Convinced of the quality of their water, and of the desirability of a moderately carbonated refresher alongside the higher fizz of many Swiss waters, the owners ingeniously decided to begin by giving their product away, a case of 20 1-litre bottles at a time, to individual households in the region of Chur.

One month later they returned with replacements asking householders to pay this time (a little above most mineral water prices) and promising regular monthly delivery. It worked; 70 per cent of households renewed the order and got to like the personal service of delivery to the home.

Gradually Valser delivered its water outside its own canton, and throughout the entire German part of Switzerland. Although supermarkets and restaurants soon asked to stock Valser, the company still retains its household service, which forms one-third of its total business.

Its loyal following has made it the number one water in the home consumption sector of the mineral water market in Switzerland. In 1982, it began to produce smaller bottles for restaurant use. Now, with its direct household deliveries, the sales in supermarkets, hotels and restaurants, Valser is the number two in the total Swiss market.

Smiling over the successful climb, managing director Max Lienhard says artlessly, 'We simply say in our advertising, "It is good, Valser water." We make no other claims.' In fact, the water is accepted on the German market as a *Heilwasser* (health water) with recognition of its effectiveness for digestive and intestinal problems, and as a diuretic

Following the popularity of Valser St Petersquelle, the spa hotel in Vals has also re-opened , offering visitors the chance to relax in the warm waters once again to sooth their rheumatism or other ailments, or simply to ease the strain of modern life.
Production: about 100 million bottles.
Exports: to Germany and the United States.

Bottling company:
Valser Mineralquellen AG,
3097 Bern-Liebefeld,
Steinhölzli,
Switzerland.

Analysis:	*mg/l*
Magnesium	54.00
Calcium	436.00
Sodium	10.70
Iron	1.83
Bicarbonates	386.50
Sulphates	980.00
Fluorides	0.63
Chlorides	2.50

Rocks from which Valser draws its minerals.

Passuger: moderately carbonated, high mineralisation.

Rhäzmünser: still and highly carbonated, high mineralization.

Allegra: still, low mineralization.

PASSUGGER

PASSUGGER

The small green bottle (a mere third of a litre), with its incised label, which arrives on your restaurant table in Switzerland, is most likely to be Passugger, number one in the Swiss catering trade. The other water, Rhäzünser, mainly sells in supermarkets, but in recent years has been offered to us in restaurants as an alternative to Passugger. In that sense the two are almost interchangeable. Between them they enjoy almost 10 per cent of the Swiss market, with Rhäzünser selling almost twice as much as Passugger. The company also bottles four medicinal *Heilwässer* (health waters) in small quantities.

The Passugger water comes from the Theophil spring in the Alpine valley of Passugg, near Chur. The water takes around 14 years to make the journey from snow and rain on the surrounding mountains and emerges as a highly mineralized water at a cool 8°C (46.4°F). The water is piped down simply by gravity to the nearby factory, and in this movement loses most of its natural carbon gas. Carbon gas from elsewhere is injected into the water in the bottling process at a level of 5 grams per litre.

The spring has been known and used since the 16th century, but bottling began only after 1850. In 1896 the Passugger family bought the spring, and at first concentrated on a hotel and spa. But after the Second World War, tourism declined, and attention was turned more seriously to bottling. In 1950, a mere 2.7 million bottles were produced. Today, production is over 150 million of the small glass bottles (the company having resisted the move to plastic). The flow of Passugger is somewhat limited at a modest 50 litres (13.2 US gallons) per minute, but Rhäzünser is a more plentiful source, which explains why, in the expanding Swiss market, its sales easily outstrip those of Passugger. The company is now also marketing a new, lightly mineralized still water called Allegra, which we first drank in our hotel in Zurich in 1994.

Production: over 55 million litres (14.5 million US gallons)

Bottling company:
Passugger Heilquellen AG,
7062 Passugg,
Switzerland.

Analysis:	*mg/l*
Sodium	*46.00*
Potassium	*3.40*
Magnesium	*24.00*
Calcium	*286.00*
Iron	*0.57*
Fluorides	*0.10*
Chlorides	*19.00*
Sulphates	*47.00*
Bicarbonates	*1,020.00*

APROZ

Aproz is the mineral water of the Migros supermarket chain in Switzerland. It is a highly mineralized water, somewhat akin to Contrex or Vittel in France, that is sold under three labels; Cristal (still), medium (with added carbon gas of 4 grams per litre) and nature (with added carbon gas of 6 grams per litre). Nature is the most popular, accounting for 87 per cent of all sales.

Aproz's history is brief. In 1947 two men from the village of Aproz on the banks of the Rhone in the Valais region of the Alps started bottling the waters of a local mountain spring in a wooden shed. They sold only 8,000 litres the first year, but persisted until 1958 when they topped 4 million (1 million US gallons) and were then bought out by Migros, under whos hand a modern plant has been installed.

Originally the water came from springs in the hillside immediately behind the factory, but the catchment area has now been extended to two other sources higher up in the mountain, including one nearly 2 kilometres away on the brow of a ridge at 1,226 metres. The blend from four springs currently in use, named Ancienne, Milieu, Pollen and Confartyr, has a particularly high sulphate content and is also rich in bicarbonates, calcium and magnesium.

Production: 70 million litres (18 million US gallons) annually.

Aproz: still and moderately or highly carbonated, high mineralization.

Bottling company:
SEBA Aproz SA,
Case Postale 815,
CH-1950 Sion,
Switzerland.

Analysis:	*mg/l*
Sodium	*8.20*
Potassium	*2.10*
Magnesium	*67.00*
Calcium	*454.00*
Strontium	*5.20*
Fluoride	*0.13*
Chlorides	*8.30*
Nitrates	*3.60*
Sulphates	*1,118.00*
Bicarbonates	*229.00*

AQUI

When Zurich's Hurlimann Brewery needed more water for its beer in 1974, a water diviner assured the president, Martin Hurlimann, that he had only to drill beneath his factory to find all he needed. Sure enough 450 metres (1,500 feet) down a giant aquifer was tapped. It proved to be an excellent mineral water, with 1,050 mgs per litre of mineral salts including sodium bicarbonate, sulphates and a nice touch of fluoride. The water was easily accepted as a beneficial health drink and was soon being marketed by Hurlimann as a mineral water in its own right, in both a still and a moderately carbonated version (5 grams per litre). The brewery also built an ultra modern fountain in front of its establishment so that the local people could drink the new water.

The company chose an ammonite as the symbol for the Aqui label to indicate that the water originates in a deep layer of miocene sandstone. Although it is below the city, the aquifer's great depth ensures that it is free from surface pollution.

Aqui is bottled in glass on Hurlimann's beer lines three days a week, and is sold as an expensive, rather prestige, water in restaurants and shops. Slightly alkaline, Aqui is billed as an aid to the digestion, and the fluoride content is sufficient to prevent tooth decay in children.

Production: 5 million litres (1,320,000 million US gallons) a year.

Exports: 10 per cent of production to the United States, Canada, Saudi Arabia and Egypt.

Aqui: still or moderately carbonated, moderate mineralization.

Bottling company:
Brauerei Hurlimann,
AG,
Brandschenkestrasse
150,
CH-8029 Zurich,
Switzerland.

Analysis:	*mg/l*
Sodium chloride	*104.0*
Sodium bicarbonate	*216.0*
Potassium	*1.8*
Calcium	*2.5*
Fluorides	*3.2*
Chlorides	*160.0*
Sulphates	*130.0*
Bicarbonates	*410.0*

AUSTRIA
INTRODUCTION

Although the origins of mineral water use in Austria go back, as in most European countries, to the Romans who exploited the hot volcanic springs that surface in many parts of the country, the modern era really began in the days of the Austro-Hungarian Empire in the 18th and 19th centuries. The Emperor Franz Josef took the waters at Bad Gastein in Styria, and Viennese high society rode to nearby Bad Vöslau, or journeyed down to the tiny spa of Bad Gleichenberg near the present border with Slovenia.

No one fostered the image of healing waters more throughout the Austro-Hungarian lands in the 19th century than an energetic fellow named Heinrich Mattoni who started bottling the waters at Karlsbad (in what is now the Czech Republic), at Hunyadi Janos near Budapest (in present-day Hungary), and at Giesshubel (now Olešnice), near the Czech-Polish border. Mattoni's imprint was almost as important as the water itself. He sold water from one of the innumerable Karlsbad springs, labelled 'Heinrich Mattoni's Karlsbader Mineralwasser'. 'Mattoni was a real emperor of mineral water,' Dr Jurgen Frank, who has assembled a unique collection of old Mattoni clay and glass bottles, told us. 'He started selling miniature bottles as toys for children, and he had his own violet bottles with HM on them for his personal consumption.'

Founder of Bad Gleichenburg, Mathias von Winkenberg.

The early days of mineral water sales in Austria centred very much on the bottling of spa waters, which were often exceptionally highly mineralized by European standards. Gleichenberger from Bad Gleichenberg, and Sicheldorfer from Bad Radkersburg, for example, both contain over 5,000 mg of minerals per litre – 10 times the amount in a French water like Evian or Volvic. Even 20 years ago the market was centred on traditional spa waters like Güssinger (where the bottling plant dates back to 1815), Gasteiner and Preblauer, whose output was often (and still is) vey modest. Indeed, going the rounds in Austria we often encountered small family operations tucked away in picturesque valleys. To get to Peterquelle, for instance, which was once just a good spring which local farmers frequented, we had to pick our way through chickens and ducks in a farmyard to reach the modern bottling factory.

Since the late 1960s, however, the picture has changed abruptly. The more traditional spa bottling companies selling highly mineralized Heilwasser (health waters) have been thrust aside in the rapidly growing market that has expanded from a mere 50 million litres (13.2 million US gallons) to over 600 million litres (158.5 million US gallons). Today the Austrians are fifth equal in the European consumption league at 76 litres (20 US gallons) per head. But it remains very much a domestic market dominated by sparkling waters, which account for well over 90 per cent of sales. Imports and exports are negligible. Although there are 18 mineral

Fresco celebrates the hot springs of Bad Vöslau.

water companies, the market is dominated by Römerquelle and Vöslauer which account for nearly half. Operating from automated modern factories, their highly competitive prices mean the consumer can buy a litre of mineral water in the supermarket, even though it is subject (over the industry's protest) to a high value added tax, for less than the cost of a daily newspaper. The major companies have promoted a healthy, sporting image for water, and encouraged the habit of mixing mineral water with white wine as a spritzer (a custom much helped by tough drinking and driving laws).

The pressure of competition has forced some of the older companies, squeezed by soaring transport costs and 20 per cent value added tax, to retrench. Güssinger, for example, now concentrates entirely on the restaurant trade where higher premiums can be earned, while Alpquell from Münster which supplies Salzburg and western Austria has shrewdly tapped a new spring, Riedquell, close to Vienna to serve the east. But others are being forced to withdraw from the national scene and supply just their local communities.

All the waters in Austria are controlled under a dual set of federal and provincial regulations, although these have been brought largely into line with those of the European Union. Basically the Federal Drug Institute licenses a mineral and health water if it contains over 1,000 mg of minerals per litre; and most waters accept the joint description (unlike the German custom of distinguishing between mineral and *Heilwasser*). Provincial legislation then obliges the bottling companies to specify on their labels both the chemical analysis and any therapeutic value in their waters. The law also still insists on glass bottles. Austrian water producers have never been able to agree entirely among themselves on a common bottle, and half still insist upon their own individual bottles. The modest scale of most operations has also meant that exports have never achieved any real significance. Everyone concentrates on the home market.

National association:
OHKV Verband,
Wiedner
Hauptstrasse 63,
A-1045 Vienna,
Austria.

Romerquelle: lightly and moderately carbonated, medium mineralization.

RÖMERQUELLE

Tucked into a fold in the flat countryside east of Vienna close to the Slovakian border you suddenly come upon the little village of Edelstal. Just over 500 people live in the houses clustered round the church, and many of them, besides tending their farms and vineyards nearby, work in the sprawling modern factory of Römerquelle, which now bottles water from a natural spring that the villagers themselves drank and bathed in for 2,000 years. Today Römerquelle (literally Roman Spring) is Austria's best-selling natural mineral water.

The Roman connection is based upon legend that the Emperor Marcus Aurelius drank the waters of the spring. Village records certainly show the artesian spring being used s for two millennia. The water comes from a deposit 400 metres (1,308 feet) deep, and is moderately mineralized with a blend of calcium, magnesium, bicarbonate and sulphate, and emerges at a constant 17°C (62°F).

Originally the villagers drew water from the spring and built a small bathhouse beside it, but Edelstal never had pretensions as a fashionable spa. And it was not until 1951 that bottling started, and the *'Heilquelle'* approval as a health water was secured only in 1962. But since 1965, when it was taken over by the present partners, Römerquelle has forged ahead, not just to dominate the life of the tiny community, but to become Austria's leading water, promoted very much with a sporting image.

To cope with such success the partners have now drilled 11 boreholes into the aquifer, of which the water from five is normally used for bottling. The water emerges under its own pressure with a very mild CO_2 content of 0.5 grams, but additional CO_2 is then added to produce the two versions marketed; one with 2 grams per litre, the other with 5 grams.

Production: 185 million bottles (about 50 million US gallons).

Bottling company:
Romerquelle Gesellschaft mbH,
Holzmanngasse 3,
1210 Vienna,
Austria.

Analysis:	*mg/l*
Sodium	12.80
Magnesium	65.00
Calcium	145.50
Chloride	4.50
Sulphate	292.30
Nitrates	0.55
Bicarbonates	424.40

Village of Edelstal, source of Römerquelle.

Vöslauer: moderately carbonated, medium mineralization.

VÖSLAUER

The warm thermal baths of Bad Vöslau on the foothills of the Harz mountains, 30 kilometres (18 miles) south of Vienna, have been visited by the Austrian aristocracy since the 18th century to drink the water to ease digestive ailments, or to bathe in it to soothe their injuries. More recently, it has been bottled and for nearly 20 years has been securely in place as the second most popular water in Austria with a healthy 18 per cent of the market. Vöslauer has carefully cultivated the image of a refreshing water, rather than simply seeking to promote its reputation as a mineral water.

The deep volcanic hot springs of Bad Vöslau rise through a natural fissure from a depth of 600 metres (2,000 feet) in an abundant flow of 3.7 million litres (980,000 US gallons) a day at a constant 24°C (75.2°F). The water blends magnesium, calcium and sulphate, but in much more moderate amounts than most of its Austrian rivals. But its natural warmth (which gave rise to a misty haze over the thermal baths on the wintry day we were there) pervades the cluster of swimming pools and remedial baths. More than 11,000 people still come to Bad Vöslau annually, many of them to ease injuries caused in accidents. The bottling company Vöslauer Heilquellen, now a wholly owned subsidiary of Central Sparkassen, the Vienna savings bank, was formed in 1936. Its water is drawn off through a stainless steel delivery pipe connected with a valve at the top of the spring (see illustration) before it erupts into the thermal baths, and is pumped 500 metres (1,600 feet) to the bottling plant. Although the water is naturally very mildly carbonated, CO_2 is added artificially to produce a moderately carbonated version at 3.5 grams per litre (which accounts for about 10 per cent of sales) and the fizzier, more popular brand with 5.5 grams. Vöslauer is distributed throughout Austria, mainly through supermarkets.

Production: over 100 million litres (26.3 million US gallons).

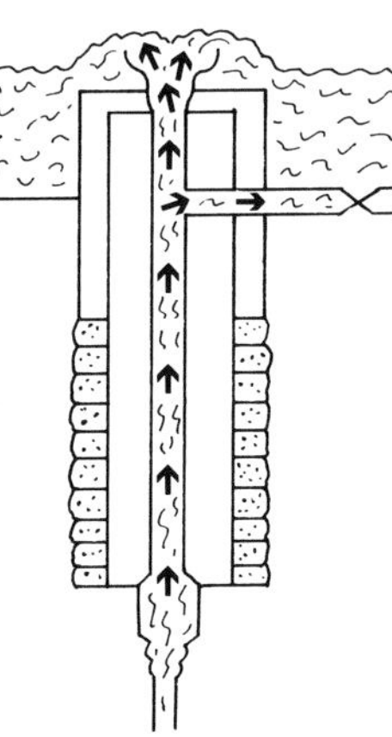

Bottling company:
Vöslauer Heilquellen GmbH,
Bahnstrasse 11,
2540 Bad Vöslau,
Austria.

Analysis:	*mg/l*
Sodium	*5.11*
Magnesium	*36.76*
Calcium	*56.63*
Chloride	*5.55*
Sulphide	*35.09*
Bicarbonate	*52.82*

Thermal baths at Bad Vöslau.

REGIONAL WATERS

From the mountains of the Tyrol in the west, to the Bohemian plain in the east, Austria offers a dramatic contrast in landscape within its small area, and a similarly wide contrast in mineral waters. Alpine springs and artesian wells exist in plenty, some going back to Roman usage. Some of the thermal waters near the Czech border and south of Vienna surface at high temperatures, while others, in southern Styria, are cool. A thriving tradition still exists for 'taking the waters' at spas all over Austria, to prevent or cure kidney stones, soothe intestinal problems, or simply to dilute the excesses of rich Viennese cooking. Austrians enjoy a spritzer in summer, half white wine and half mineral water.

Alpquell, Riedquell and **Astoria** all belong to the Quellen-Betriebe group of Richard Rieder, which ranks fourth in Austria with 10 per cent of the market. Rieder first moved into mineral waters in 1972 when he bought Alpquell. Alpquell comes from the Tyrol mountains near Münster and he was soon distributing it nationwide. Then he acquired Astoria, a light water from the Bohemian plain, targeting it as a discount water in supermarkets. Finally he launched Riedquell from a bottling plant at Shrick, also on the Bohemian plain, close to the Czech border. Since transport costs in Austria are high – most water is still bottled in communal glass bottles as in Germany – this had the Quellen-Betriebe group strategically placed with a source near all major population centres. Both Alpquell and Riedquell are similar in composition, being high in calcium, magnesium and bicarbonates; iron is removed from both, and carbon gas added. Production of the group is around 60 million litres (15.8 million US gallons).

Three other waters, **Gasteiner**, **Güssinger** and **Severin Quelle** are owned by the group, Brau AG, and, combined, rank third in Austrian sales with a 12 per cent share.

Gasteiner is bottled in the spa of Bad Gastein in south west Austria, which was in fashion in the 19th century after the Emperor Franz Josef decided to take the hot baths there. The present-day bottling factory produces about 20 million litres (5 million US gallons) from the artesian well, a spring rising at 40°C (104°F) but relatively low in minerals.

Güssinger in southern Austria produces a highly mineralized water that was known to the Romans and has been on record since 1469. The water was famous among the nobility of the Austro-Hungarian Empire as a classic aid to intestinal functions, and drunk at the fashionable local spa. Bottling started in 1956. The mineral water rises at 14°C (57°F) from an artesian well and is carbonated for the market at two levels of gas, with 6.5 grams per litre or a milder 2.5 grams per litre, both selling around 8 million litres (2 million US gallons) a year. Güssinger concentrates on the restaurant market (where it runs second to Vöslauer) and was the mineral water served automatically to us in the bar of Vienna's famous state opera. However, the high level of minerals in Güssinger makes it less suitable to use as a spritzer mix with white wine. Production now runs at about 22 million litres (6 million US gallons), with some exports to the United States and Australia.

Severin Quelle belongs to the same company as Güssinger, and was first bottled in 1974. The water is less highly mineralized than Güssinger, but even so recommended for kidney stones and the digestion.

Gleichenberger from Bad Gleichenberg in south-eastern Austria, 210 kilometres (130 miles) from Vienna, was rescued from oblivion by Mathias von Winckenberg, governor of Styria, in the 19th century. The waters had been known to the Romans but fallen into disuse until the spa was opened in 1836. From then on fashionable Austrians came to take the three local waters which, along with the mild local air, were recommended for bronchitis and asthma as well as circulatory problems and digestive ailments. Highly mineralized (over 5,000 mg of mineral salts per litre), with a high sodium bicarbonate level, numerous trace elements and natural carbon gas, Gleichenberger nevertheless is now bottled in small quantities, 2 or 3 million litres (500,000 to 800,000 US gallons), for regional use, although the spa, its facilities, park, and graceful 19th-century buildings attract several thousand visitors in the summer months.

Peterquelle in the region of Deutsch Goritz, in south eastern Austria, was merely a spring in the local woods used for nearly two centuries by farmers until Frau Schweiger built a factory in 1960 and began to sell the mineral water throughout Austria. She quickly built up a strong demand, with publicity emphasizing the sporting and health benefits of her water, and pictures of motor racing champions like Niki Lauda quaffing Peterquelle. With total mineral salts of 3,457 mg per litre, the water is highly mineralized by international standards, and particularly high in sodium bicarbonates and calcium. Two artesian wells have been drilled and the water is pumped a mere 300 metres (984 feet) into the modern factory, surrounded by a picturesque farm. Production is about 30 million litres (8 million US gallons).

Sicheldorfer, which comes from south eastern Austria, not far from the Slovenian border, was discovered when a local company was boring for oil in 1923 and discovered mineral water instead. The spring contains a very highly mineralized water with 5,400 mgs per litre of mineral salts, high in sodium bicarbonates and containing numerous trace elements. The water rises through basalt rocks in what was a volcanic region at a temperature of 11.5°C (52.7°F) from a borehole 55 metres (180 feet) deep in an artesian well. In 1956 the bottling factory was built, and it is the only industry in the small village of Bad Radkersberg (population 150). The water sells with 3.6 grams per litre of carbon gas. Locals arrive and collect their weekly supply from the factory, although it is distributed as far as Vienna. Sicheldorfer claims to be the most highly mineralized water in Austria. Production is about 10 million litres (2.6 million US gallons).

UNITED KINGDOM
INTRODUCTION

Original label idea for new U.K. water.

The beer-loving British now spend £1 million ($1.5 million) a day on bottled water – a sure indication that the water habit is here to stay. In spite of economic recession and a series of poor summers, bottled water is the fastest growing sector of the UK food and drinks market. In 1993, people in the UK drank 570 million litres (150 million US gallons) – three times as much as they swallowed five years ago, and 10 times more than they downed in 1983.

'Wealth, health, travel, and concern about tap water have fuelled demand in the UK for bottled water,' says Richard Hall, chairman of market research company Zenith International. He pinpoints the two most significant events in the development of the UK market: Perrier's brilliant advertising campaigns of the 1970s and 1980s – Remember 'Eau La La!' and 'H_2Eau'? – which created a desirable image for water in Britain; and the company's recall of its stocks from British supermarket shelves in 1990 when there was a possibility that they might have been contaminated by benzene. 'This', he says, 'allowed other, mainly British producers to gain a foothold in the market.'

There are now more than 200 bottled waters, including imported brands, available in Britain, and a dozen or so newcomers appear on the market each year. The overall favourite is Évian, owned by the French food and beverage giant, Danone (formerly BSN), which has an 11 per cent share of the British market. But imported waters have become less popular over the past five years, and now account for only one third of British consumption, compared with more than half in 1988. During that time there has been a big swing in demand towards home-produced waters.

Around 60 British brands are natural mineral waters, registered under the UK and European regulations, a small proportion are purified tap and other water, and the rest are mainly spring waters, bottled to high standards of hygiene. The two largest British producers are Highland Spring and Campsie Spring Scotland, which between them account for nearly a third of total output and 85 per cent of supermarket own-label production.

Other major players in the UK market are Perrier (UK), part of Nestlé Sources International, the Swiss conglomerate, which owns those two most English of sources, Ashbourne and Buxton (the top-selling British water); Strathmore, whose carbonated spring water is nudging Perrier for the lead in the sparkling stakes; and Eden Valley, which successfully launched Aqua-Pura with a Roman image.

The top 10 UK companies, all of which have less than a 20 per cent share of the market, produce 78 per cent of Britain's bottled water, and there are dozens of even smaller producers. Just as the French have their local cheeses, the British have their regional waters, many of which are developed by farmers and landowners as 'value added' rather than full-time enterprises.

Almost every region in the UK can boast its own spring or mineral water bottled at source. Some, like Somerley, are distributed nationally, and, given enough investment and publicity, may become market leaders. Others, such as Blenheim, remain in their own locality, and are supplied to hotels, restaurants and grocery stores at competitive prices. Yet others, like Decantae, Tŷ Nant and Fionnar, maintain a relatively small following at home, but sell well abroad, often in carefully selected niche markets.

For many Britons, the bottled water habit has become part of everyday life. Most good hotels and restaurants routinely offer bottled water as an accompaniment to food. Some even have a water as well as a wine list. When we visited the Bath Spa Hotel, for example, there was a choice of 14 waters ranging from well-known French and Italian brands to Thirsty Camel, an "easily quaffable" sparkling English spring water.

Selected brands are not only offered by winebars, cafes and restaurants, but also "adopted" by magazines, fashion houses, conference centres, football teams and rugby squads. Producers sponsor sporting events like the London Marathon, and at least three charities boost their funds by selling their own brands. Even babies will soon have their own bottled water, produced by Gleneagles for H J Heinz.

While supermarkets are by far the biggest stockists, bottled water is also available in petrol stations, newsagents, chemists, fast food outlets, schools, colleges, dairies, sports clubs, garden centres and corner shops – places where, five years ago, you would never expect to see it. Doorstep delivery of water, often along with the milk, accounts for an annual 13 million litres (3.4 million US gallons). Water coolers in offices, factories and other workplaces dispense another 38 million litres (10 million US gallons) a year – 7 per cent of total UK bottled water consumption.

Several other developments have taken place in recent years. There has been a big swing from carbonated waters like Perrier, which dominated the market up to the end of the 1980s, to the still varieties, which now form 60 per cent of UK sales. There has also been a considerable rise in demand for spring waters, which now account for more than a fifth of total UK consumption. Sales of flavoured waters – the most popular being lemon, lime and orange – are growing at an annual rate of 18.5 per cent, and more exotic varieties, like elderflower and wild blackberry, are on the shelves or in the pipeline.

The British market is volatile and intensely competitive. 'UK bottled water companies', says Richard Hall, 'are at the forefront of innovation – in bottle and label design, packaging, marketing and the types of water they produce.'

National Association:
Natural Mineral Water Association,
20-22 Stukeley Street,
London WC2B 5LR.

Princess Diana has obviously acquired the mineral water habit.

Abbey Well: still and highly carbonated, low mineralization.

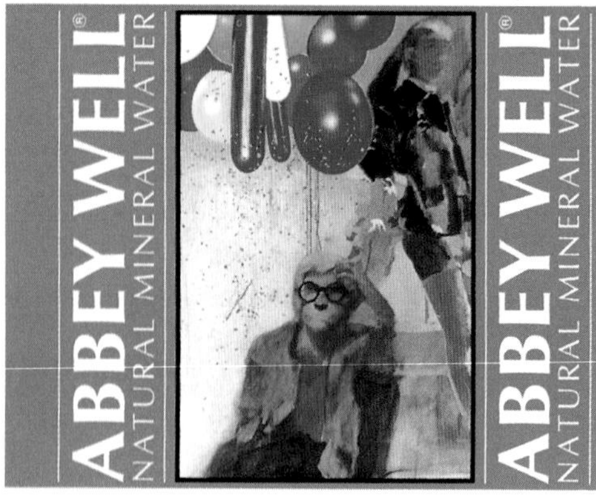

ABBEY WELL

Abbey Well is easily recognisable as it is the only brand of bottled water to feature a pop art label – a colourful portrait of David Hockney by fellow artist Peter Blake.

This clean tasting water, low in sodium, calcium, nitrate and other minerals, is produced by Waters & Robson Ltd, a privately owned company which has been manufacturing soft drinks in the Northumbrian town of Morpeth since 1910. It was Thomas Robson, grandfather of the present managing director, Tony Robson, who sank the company's first well head in that year, and named it Abbey Well after his favourite local landmark, a 12th century Cistercian abbey.

In 1982 Abbey Well made its debut in the UK as a bottled water, and four years later it was officially recognised as a natural mineral water under the National Mineral Water Regulations. It is now one of the top 10 brands of water produced in Britain, and provides work for 240 employees. The water is taken from the original source 118 metres (387 feet) beneath the Northumbrian hillside where it percolates through white, water-bearing sandstone deposited more than 300 million years ago. It is brought from the well head by stainless steel piping to the bottling plant where there are at present three bottling lines.

Abbey Well is distributed nationally, and is especially popular in London and the south east of England. You won't, however, find it on your supermarket shelves. The company has a strict policy of not supplying the large grocery multiples to ensure that its stockists have the benefit of better profit margins and consistent supplies. Instead Abbey Well is distributed through independent and specialist retailers, and through breweries, wholesalers and dairies.

It has a 14 per cent share of the 12-million-litre doorstep delivery market, and is the leading brand of British mineral water supplied to the UK licensed trade, being stocked in both still and sparkling versions by many of the country's best-known hotels and restaurants.

Production: nearly 14 million litres (3.7 million US gallons).

Exports: about 1 per cent of production.

Bottling company:
Waters & Robson Ltd,
Abbey Well,
Morpeth,
Northumberland
NE61 6JF.

Analysis	*mg/l*
Calcium	*54.00*
Magnesium	*36.00*
Potassium	*7.50*
Sodium	*45.00*
Carbonate	*173.00*
Chloride	*80.00*
Fluoride	*0.09*
Nitrate	*0.90*
Sulphate	*28.00*

Aqua-Pura: still and highly carbonated, low mineralization.

AQUA-PURA

A moderately hard, clean-tasting natural spring water from Cumbria, Aqua-Pura was launched in October 1991. It soon ranked fifth among UK-produced bottled waters as a result of selective marketing, competitive pricing and distinctive labelling, Aqua-Pura was discovered during a countrywide survey commissioned by the Robinson Group, owner of the Stretton Hills brand, and carried out by the Water Research Council. Asked to find a suitable site for development, independent hydrologists recommended the unspoilt, virtually unpopulated Eden Valley on the edge of the English Lake District. Ninety metres (295 feet) beneath the Penrith sandstone and shales, protected from possible contaminants by a layer of boulder clay, lies an estimated 3 billion litres (792 million US gallons) of pure, lightly mineralized water.

This is now extracted and bottled at source on the company's greenfield site near the village of Armathwaite. As that part of Britain, being near Hadrian's Wall, has strong Roman connections, the company decided to name its new water Aqua-Pura, and to market it with the picture of a Roman water jar on the label.

Although Aqua-Pura meets all the official requirements for natural mineral water status, it is not registered as a source. The company prefers to market it as a spring water. It is bottled to the standard of European directive 777, which guarantees that the water comes from an approved source and is free from pollution and toxic substances. Both still and sparkling versions are available, the latter having 8 grams per litre of carbon gas added during the bottling process.

Aqua-Pura has been marketed successfully as a medium-priced product for supermarket customers who buy large amounts of bottled water for everyday consumption, but who prefer a premium to an own-label brand. 'Aqua-Pura's stronghold is in the grocery multiples, says Marketing Manager Heidi Lucas. 'During 1993', she adds, 'Aqua-Pura outsold every brand in the supermarkets'.
Production: 20 million litres (5.3 million US gallons).
Exports: small amounts on an ad hoc basis to Russia, Cyprus and the Middle East.

Bottling company:
The Eden Valley Mineral Water Company Ltd, Armathwaite, Cumbria CA4 9TU.
Owned: Robinson Group of Companies Ltd.

Analysis:	*mg/l*
Calcium	*52.5*
Magnesium	*7.0*
Sodium	*26.5*
Potassium	*2.5*
Chloride	*80.0*
Sulphate	*17.5*
Nitrate	*24.0*
Fluoride	*< 0.1*

Brecon Carreg: still and highly carbonated, low mineralization.

Rock Spring: still, low mineralization.

Bottling company
Spadel Ltd,
2 Clifton Court,
Corner Hall,
Hemel Hempstead,
Herts HP3 9XY.
Owned: Spadel.

Brecon Carreg:	*mg/l*
Calcium	*47.5*
Magnesium	*16.5*
Sodium	*5.7*
Potassium	*0.4*
Chloride	*13.0*
Sulphates	*9.0*
Nitrates	*2.2*
Hydrocarbonates	*206.0*
Silica	*5.1*

Rock Spring:	*3.2*
Sodium	*5.0*
Potassium	*0.4*
Chloride	*12.0*
Sulphates	*10.0*
Nitrates	*1.3*
Hydrocarbonates	*67.0*

BRECON CARREG

The natural mineral water springs of Brecon Carreg and Rock Spring rise from an aquifer 63 metres (240 feet) beneath the Brecon Beacons National Park in the mountains of south Wales. The area, which covers more than 129,500 hectares (500 square miles), is protected by law from industrial, urban and agricultural pollutants. The springs themselves are located in what were the estates of the Earl of Cawdor, and have supplied water to local residents since the end of the last century.

In 1979, a small company was set up to bottle and sell Brecon waters locally. Four years later it was purchased by the Belgian conglomerate, Spadel, producer of the well-known Spa and Monastère brands of mineral water, which are also sold in Britain. The new owner started to expand the business into a major commercial operation, investing more than £5 million in building a modern bottling plant, which was designed to blend in with the surrounding area. More than 3,000 trees were planted to enhance the landscape.

The Brecon waters owe their purity to the layers of millstone, grit and sandstone through which they slowly percolate. Having reached the insoluble band of limestone which lies beneath, they then flow rapidly underground. Because contact with the limestone is relatively brief, their mineral and salt content remains low. They are both fresh, clean-tasting waters, Rock Spring having the lower mineral content of the two.

The waters are bottled at source, and subjected to hourly quality assurance checks. Brecon Carreg is sold in still and sparkling varieties, the carbonation being added during the bottling process. Rock Spring is available only as a still water. Both waters are bottled only in PET, and can be found in supermarkets and grocery stores throughout Britain. Rock spring has even graced the shelves of Harrods.

Production: Brecon Carreg 6.1 million litres (1.61 million US gallons); and Rock Spring 1 million litres (0.264 million US gallons).

Buxton: still and highly carbonated, low mineralization.

BUXTON

Buxton is a light, pure water from the beautiful Derbyshire Peak District. The clear, mountain water rises from a subterranean reservoir over 1,500 metres (4,900 feet) deep. The underground journey filters the water naturally, a process that geologists calculate takes up to 6,000 years. Buxton has full natural mineral water status complying with the stringent European Union and UK regulations.

The Romans were the first to appreciate the warm Buxton waters, naming their settlement Aquae Arnemetiae, or the Spa of the Grove Goddess. Medieval pilgrims took the waters, as did Mary, Queen of Scots, who visited Buxton to cure her rheumatism between 1570 and 1584 when in captivity nearby. But the 18th century was the fashionable heyday when the fifth Duke of Devonshire built the Crescent to house hotels and assembly rooms for visitors bathing in and drinking the warm waters. No thermal treatments are available today, however.

The water is tapped as it rises through a rock fissure in what was, in Victorian times, the Gentlemen's Thermal Baths. Today the pool is canopied and covered with sterile air, from which two pipes draw off the water, one to St Ann's Well, the public drinking fountain, and the other to the modern bottling plant, which is capable of filling up to 8,000 bottles an hour.

The water at Buxton emerges at a rate of 768,000 litres per day at a constant temperature of 27.5°C (81.5°F). It carries a predominantly calcium content, which is considered to be of therapeutic benefit. An average of 6.5 grams per litre of carbon gas is added in the bottling process to produce the sparkle in the carbonated version, but there is an increasing demand for the still water.

Since it was acquired by Perrier (UK) Ltd, now part of Nestlé Sources International, the Swiss food and beverages giant, Buxton has become the leading brand of bottled water produced in the United Kingdom, accounting for nearly 7 per cent of total output. The company also has a water cooler division, formed six years ago to service offices, factories and department stores, and has introduced a range of flavoured mineral waters.

Production: 45 million litres (11.9 million US gallons).

Public drinking fountain today at Buxton Spa.

Bottling company:
Buxton Mineral Water Company Ltd,
Trinity Court,
Church Street,
Rickmansworth, Herts
WD3 1LD.
Owned: Nestlé Sources International.

Analysis:	*mg/l*
Calcium	55.0
Magnesium	19.0
Sodium	24.0
Potassium	1.0
Bicarbonates	248.0
Chloride	42.0
Sulphates	23.0
Nitrates	0.1

Chiltern Hills: still and moderately carbonated, low mineralization.

CHILTERN HILLS

Chiltern Hills was the first English water to be recognised as a natural mineral water under the UK regulations which came into operation in 1985. The source of this lightly mineralized, clean tasting water is at Toms Hill Estate, near the village of Aldbury in west Hertfordshire. The private estate, home for many years of Denis Ward, chairman and owner of J W Ward & Son Ltd, the company which bottles and markets Chiltern Hills, is within a designated area of outstanding natural beauty adjoining the Ashridge National Trust property.

The water is extracted from an aquifer 100 metres (350 feet) deep in the 95-million-year-old lower chalk strata of the Chiltern Hills. It originates as rainwater which takes at least 50 years to percolate through the layers of pure limestone to the aquifer below. During this filtration process, the water is purified, and absorbs a certain amount of calcium and trace elements. The calcium content is in line with the guide level specified in the EU drinking water directive. Chiltern Hills is low in sodium and nitrates, and contains a small amount of naturally occurring fluoride which is helpful in preventing tooth decay.

After his family timber business closed in the early 1980s, Denis Ward decided to develop the water source on his estate. Analysis confirmed that the water was of excellent quality, and in 1983 a small bottling line was installed to exploit the source on a commercial basis.

Contracts with supermarket groups followed, and as annual sales increased, the company upgraded its production facilities. In 1991 the original dual bottling equipment was replaced by two separate new automatic bottling lines, one for still and one for carbonated water. At about the same time another line was also installed to fill 22-litre bottles for use with water coolers, which a subsidiary company supplies to offices in London and the south east of England. A further bottling line for 1-litre glass bottles was opened in 1993.

Chiltern Hills ranks among the top 10 mineral water brands in the UK, and is exported to Malta, Antigua, Ghana, Bermuda, Greece and to Cyprus.

Production: 18 million litres (4.75 million US gallons)

Exports: about 4 per cent of production.

Bottling company:
J W Ward & Son Ltd,
Bourne End Mills,
Hemel Hempstead,
Herts HP1 2RW.

Analysis:	*mg/l*
Calcium	*104.00*
Magnesium	*1.40*
Sodium	*8.00*
Potassium	*< 1.00*
Bicarbonates	*293.00*
Chloride	*15.00*
Sulphate	*12.00*
Nitrate	*5.00*
Fluoride	*0.10*
Iron	*0.02*

Highland Spring: still, lightly and moderately carbonated, very low mineralization.

HIGHLAND SPRING

Highland Spring is the largest UK producer of natural mineral water in Britain, having a 12 per cent share of the bottled water market. This light, fresh and clean-tasting water comes from the Ochil Hills in Perthshire, Scotland where it can take up to 30 years for rainfall to enter the soil, and make its way through to the aquifer. The 400-million-year-old rocks give up just enough of their minerals to create a water which has a distinctive flavour yet is low in salts.

Just over half the water the company produces comes from the Highland Spring borehole near the village of Blackford, off the Stirling to Perth highway. This is sold under the Highland Spring brand name. The rest comes from other boreholes in the same catchment area, and is bottled under the names of Mountain Spring and Glendale Spring for the Tesco and CWS supermarket chains respectively. Each natural mineral water has a slightly different chemical analysis, while retaining the overall purity and low mineralization characteristic of the premium brand.

The underground reservoir, from which Highland Spring and its sister brands are sourced, is claimed to be limitless in its capacity.

The boreholes are capped on the hillside, and secured inside a small building, where they are covered with a cement cone, from which steel pipes lead the water half a mile downhill to the extensively modernised bottling plant, where each brand is bottled independently.

From the outset, Highland Spring has proved to be a company with marketing flair. A major long-term advertising campaign has helped the water to sell well in every trade sector, from the grocery and dairy industries to catering and licensed outlets. The company was a pioneer of doorstep delivery, and is now brand leader with more than a 30 per cent share in the dairy market.

Production: 68.5 million litres (18 million US gallons) in 1993.

Exports: 5 per cent of total production in 1993 to more than 30 countries.

Bottling company:
Highland Spring Ltd,
Blackford,
Perthshire,
Scotland.

Major shareholders:
Tay Consult
Establishment,
Liechtenstein.

Analysis:	*mg/l*
Calcium	39.00
Magnesium	15.00
Fluoride	0.10
Sodium	9.00
Sulphate	9.00
Chloride	15.00
Nitrate	1.50
Bicarbonate	190.00

Malvern: still and moderately carbonated, low mineralization.

MALVERN

Malvern, the natural mineral water from the Malvern Hills in the west of England, has a simplicity and lack of taste that have long made it ideal as a mixer to bring out the best flavour in Scotch whisky. It is still especially noticeable in wine bars, hotels and high class restaurants around Britain, although it is also stocked in the major supermarkets and off-licence chains.

The many springs of the granite Malvern Hills, that form the border between Worcestershire and Herefordshire, have long been renowned for their waters. As early as 1620 the Malvern song mentioned that a thousand bottles were filled weekly 'for stomachs sickly', and dispatched as far afield as London and Berwick. The village of Great Malvern even became a spa under the auspices of a local doctor, John Wall. Then in 1851, having won the contract to supply the Great Exhibition with non-alcoholic drinks, J Schweppes & Company took over the production and sale of Malvern water.

Nearly 40 years later, Schweppes established a new bottling plant at the nearby village of Colwall, and piped to it the waters of the Primeswell Spring, which emerges at a cool 8.6°C (47.5°F) through the fractured granite of the Herefordshire Beacon, just below the site of an Iron Age fortress. The catchment area for the spring lies in the Malvern Hills conservation area, and is protected from contamination by restrictions on building, quarrying and agriculture.

The water, which flows from three crevices in the bare rock, collects in a small pool from where it is piped to tiled storage tanks close by. In 1993, the whole site was refurbished, the spring has been totally enclosed with a stainless steel cover, a new building erected to protect the installation from contamination, and a new pipeline laid, through which the water descends by gravity to the filling plant at Colwall, 2 kilometres (1.5 miles away). There the water is filtered and passed through ultra violet light to eliminate harmful bacteria. Its mineralization is very low, and it contains only modest amounts of calcium and magnesium.

Malvern enjoys the cachet that it is supplied to the Queen, who regularly takes it on tours abroad; to the Prime Minister at 10 Downing Street; and to the Houses of Parliament.

Production: 13 million litres (3.4 million US gallons).

Bottling company:
Coca-Cola & Schweppes Beverages Ltd,
Charter Place,
Vine Street,
Uxbridge, Middlesex
UB8 1EZ.
Owned: Cadbury Schweppes plc.

Analysis:	*mg/l*
Calcium carbonate	*83.0*
Magnesium carbonate	*15.0*
Magnesium sulphate	*41.0*
Magnesium chloride	*37.0*
Sodium chloride	*34.0*
Sodium nitrate	*6.0*
Potassium nitrate	*4.0*
Silica	*10.0*

Strathmore: still and highly carbonated, low mineralization.

STRATHMORE

Currently Scotland's leading brand of bottled water, Strathmore aims to become number one in the whole of the United Kingdom. The water takes its name from the Vale of Strathmore, in which lies its source, the Strathmore Springs. This unspoilt area of Scotland, just north of the Firth of Tay, has strong links with royalty. Its focal point is Glamis Castle, seat of the Earls of Strathmore since the 14th century, and childhood home of Queen Elizabeth The Queen Mother. The Strathmore labels carry a picture of the castle.

Strathmore is a calcium-chloride water with a low mineral content. Its natural hardness and purity makes it an excellent thirst quencher. The water is pumped to the surface at a constant temperature of 10.5°C (50.9°F) through a single borehole 138 metres (450 feet) deep direct to the filling plant, where there are two bottling lines, one for PET and one for glass. The sparkling version of Strathmore has between 7 and 7.5 grams per litre of carbon dioxide added during bottling.

The water was first bottled and marketed in 1984, and achieved official mineral water status a year later. It became popular in Scotland, then penetrated the markets south of the Border. In the early 1990s, sales have almost tripled, making Strathmore the fourth largest UK bottled water brand. It ranks second only to Perrier in the British market for carbonated water, and shares with Highland Spring the distinction of being the UK's top export brand.

Strathmore flavoured waters, which include lemon, lime and orange, were first introduced in 1986, initially for export to the US, but have subsequently become so popular in Britain that the company is now the UK brand leader in this fast-growing sector of the market.

Strathmore repositioned in the market as a spring water in 1992, shortly before it was acquired by Matthew Clark plc. Twenty per cent of total sales are to hotels, restaurants and other catering establishments, while 80 per cent are to retail outlets, which include everything from the Tower of London shop to the major supermarkets.

Production: 32 million litres (8.45 million US gallons).

Exports: almost 8 per cent of total production to 20 countries.

Bottling company:
Strathmore Mineral Water Company,
126 West High Street,
Forfar,
Angus DD8 1BP,
Scotland.
Owned: Matthew Clark plc.

Analysis:	*mg/l*
Calcium	*60.0*
Chloride	*125.0*
Magnesium	*15.0*
Sodium	*46.0*
Potassium	*2.2*
Carbonate	*145.0*
Sulphate	*1.0*
Nitrate	*5.0*
Fluoride	*0.1*

Caledonian Spring: still and highly carbonated, low mineralization.

Fountain Head: still and highly carbonated, low mineralization.

Gleneagles Crystal Clear: still, low mineralization.

Gleneagles Sparkling: moderately carbonated, low mineralization.

CALEDONIAN SPRING

In the Campsie Fells, high above the Scottish village of Lennoxtown in Stirlingshire, no less than nine springs are being tapped. Several of the waters are marketed as a range of own-label brands by Campsie Spring Scotland Ltd, the UK's second largest bottled water producer. So if you shop in Sainsbury's supermarkets, you will encounter Caledonian Spring; in Safeway, Glencairn Spring; or in Somerfield, Strathglen Spring. The company also produces Glenburn Spring.for Asda, and Lowland Glen for Booker. Although each water comes from a different spring, all are low in sodium and nitrates. Together they account for almost half own-label sales in Britain.

With its refreshing, slightly earthy taste, Caledonian Spring is the brand leader among these. It ranks fifth in the total British bottled water market, not too far behind long-established favourites like Evian, Buxton, Volvic and Highland Spring.

Production: 30 million litres (nearly 8 million US gallons).

FOUNTAIN HEAD

Fountain Head is a spring water with its source in the the Rough Rock aquifers under the Pennine hills to the west of Huddersfield in Yorkshire. According to scientific estimates, the rainwater which falls on the hills takes 60 years to filter through to the aquifers beneath. The water is drawn to the surface through a borehole 60 metres (200 feet) deep, and bottled at source in still and carbonated varieties by a local company, Benjamin Shaw and Sons Ltd.

Launched in 1993, the water notched up first-year sales of 19.5 million litres (5 million US gallons), and is expected to more than double this in 1994. It is sold as a premium brand under the Fountain Head label, and is bottled for Tesco, Asda and Sainsbury as Pennine Spring, Aqua Spring and Sainsbury Spring, respectively.

Production: 19.5 million litres (5 million US gallons).

GLENEAGLES

Gleneagles, a natural spring water in both still and carbonated varieties, will be making its appearance throughout the UK from January 1995. You may have already seen it in the south of England. Gleneagles Spring Waters Company, a subsidiary of Allied-Lyons, is based in Blackford, Perthshire, and draws its water from the 6,500-acre (2,631-hectare) Gleneagles Valley.

The company has two separate sources for its still and carbonated water. Both have low levels of sodium and nitrates. Gleneagles Crystal Clear is drawn from St Mungo's Spring, a volcanic source high up the valley. Gleneagles Sparkling, the carbonated version, is sourced from Castle Spring, lower down the valley, where sandstone predominates.

Four additional boreholes supply water for own-label customers, including H J Heinz & Company, whose product is claimed to be the first spring water pure enough for babies.

LOVAT

Launched in 1992, Lovat is a natural mineral water from one of the three springs which bubble to the surface at Fountainhead, near Beaufort Castle in the Scottish Highlands, family home of Lord Lovat and seat of the Clan Fraser. The other two are Glen Mor Spring, a Sainsbury own-label brand, and Lovat Spring, bottled for Iceland Frozen Foods.

Among the purest waters in Europe, they are filtered through deep layers of sandstone. Each of the waters has a slightly different mineral content, but all are low in sodium and nitrates. They are bottled at source, in both still and sparkling versions. In 1993, sales of the Lovat waters totalled 14 million litres, putting LMWS among the top 10 UK producers.

Production: 14 million litres (3.70 million US gallons) for the whole company.

NORTHUMBRIAN SPRING

The source of Northumbrian Spring natural mineral water is a huge aquifer 66 metres (215 feet) below Merrybent, in the heart of Northumbria. Northumbrian Spring is filtered through magnesium limestone, protected by a thick layer of impermeable clay, which guarantees its purity. Its subtle, yet distinctive flavour, is due to the natural mix of mineral salts it acquires on its journey to the source. It is especially suitable for low sodium diets.

For many years, Northumbrian Spring was available in catering and retail outlets, but since the production company was taken over by Water Coolers of St Albans in 1993, it supplies only water cooler units. With annual sales of 5 million litres (1.32 million US gallons), Northumbrian Spring is now one of the four leading water cooler brands in a market which is growing at an annual rate of more than 25 per cent. Northumbrian Spring's own sales in this sector increased by nearly 80 per cent in 1993.

Production: 5 million litres (1.32 million US gallons).

STRETTON HILLS

Since Saxon times, the waters of Church Stretton in Shropshire have been famous for their purity and mildly therapeutic properties, but it wasn't until the 1880s that the town's springs were commercially exploited by two soft drinks manufacturers, one of which, the Stretton Hills Mineral Water Company, was eventually acquired by Wells Soft Drinks Ltd, current producer of the Stretton Hills brand.

Stretton Hills, a pure, fresh water, is ideal for those on a low sodium diet. It gained recognition as a natural mineral water in 1987, and is now among the 10 best-selling brands produced in the UK. Sales were well over 15 million litres (nearly 4 million US gallons) in 1993, and the brand has nearly a 4 per cent share of the UK mineral water market. Bottled at source in Church Stretton, Stretton Hills is sold in still and sparkling varieties. The latter is the more popular of the two, and represents 57 per cent of the brand's sales.

Production: 15.2 million litres (4 million US gallons).

Lovat: still and moderately carbonated, low mineralization.

Northumbrian Spring: still, low mineralization.

Stretton Hills: still and lightly, moderately and highly carbonated, low mineralization.

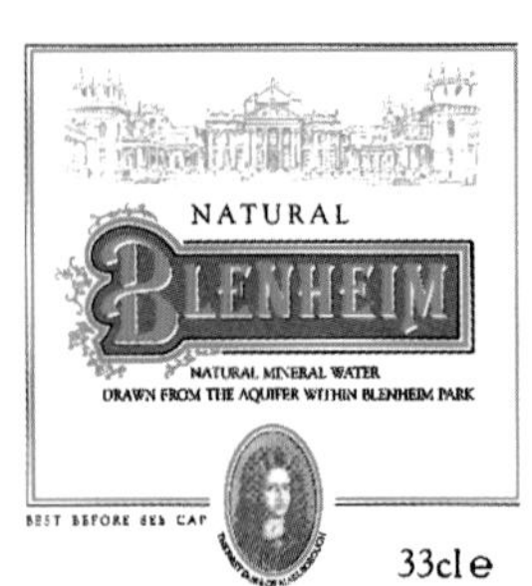

REGIONAL WATERS

From the granite of Scotland's highlands to the chalk of Hampshire's downs, the United Kingdom offers a considerable range of water sources and brands in a relatively small geographical area. Here is a selection of some of Britain's favourite regional waters.

Blenheim is a natural mineral water from a spring in the grounds of Blenheim Palace, the Duke of Marlborough's estate near Woodstock, Oxfordshire. It is drawn from Fair Rosamund's Well, known for the purity of its water, and named after one of King Henry II's mistresses who often used it as a place for bathing. There are still and sparkling versions of Blenheim, which are sold mainly to tourists and to hotels and restaurants in the area.

Caithness Spring, with its very low mineral content, is claimed to be the purest natural mineral water bottled in the United Kingdom. Its subtle, sweetish taste makes it ideal for brewing delicate teas and exotic coffees as well as for drinking on its own. The source rises half way up Scaraben, a remote Scottish mountain, 40 miles from John o' Groats. Caithness Spring was first bottled in 1984 in a custom-built plant sited at the source. Still and sparkling versions are sold in selected supermarkets and other stores under the main brand name, and the water is also supplied to British Rail and other enterprises under their own labels.

Cotswold Spring Water, drawn from the Dodington Spring, a few miles to the north of Bath, is a clear, refreshing water, naturally filtered through the oolitic limestone of the region. Being low in sodium and nitrates, it is ideal for babies and for people with heart complaints and high blood pressure. The business was founded by John and Joan Marshall in 1986, and now has 20 staff and three bottling lines. The water, available in still and carbonated versions, is bottled at source, and is distributed by the company's own vehicles throughout the west of England.

Decantae, bottled at source in the foothills of Snowdonia, was the first recognised natural mineral water in North Wales. Named after the ancient Celtic tribe which once inhabited the site of the spring, Decantae is filtered through miles of 400-million-year-old rock strata before it bubbles to the surface, making it exceptionally pure. It claims to have one of the lowest mineral contents in the bottled water market. Launched in 1985, Decantae is sold in still, sparkling and flavoured versions to selected food stores, like Fortnum & Mason, and high-class hotels and restaurants in Britain and overseas.

Devon Dew claims to be the first spring water produced commercially in its home county. Since 1985 it has been bottled at source by a family-run enterprise in Ivybridge, south Devon Output of the still and gently effervescent water is now more than 1 million litres (0.264 million US gallons) a year. It is sold mainly to the local hotel, restaurant and tourist trade. The Devon Dew brand name is restricted to the south west of England, but the water is also bottled for own-label clients, most of whom are in the London area.

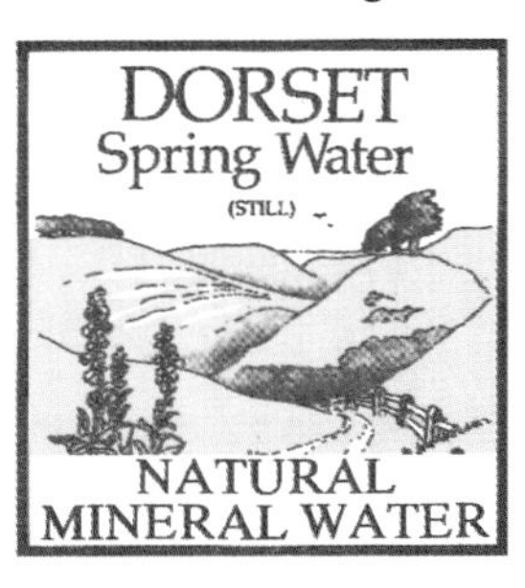

Dorset Spring Water is a still water from Old Coombe Spring near Stoke Abbott, in the beautiful farmlands of west Dorset. It was first marketed in 1980, and received recognition as a natural mineral water in 1987. Suitable for people on a low sodium diet, Dorset Spring Water is sold in health food shops, specialist grocery stores and delicatessens throughout the south of England.

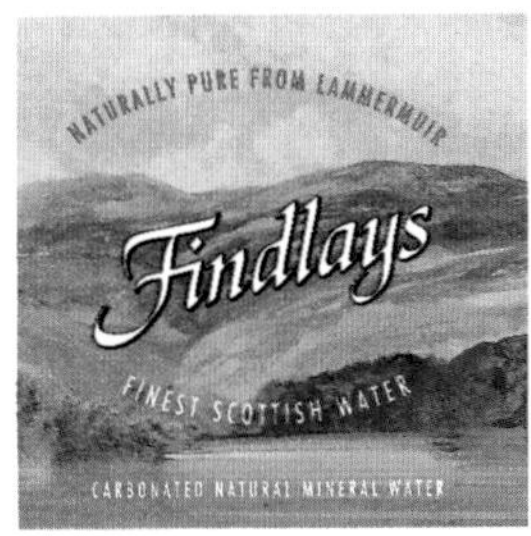

Findlays, a soft, sweet-tasting, natural mineral water, was launched in 1993. Its source is Pressmennan Well, a natural spring that rises vertically through a geological fault at the foot of the Lammermuir Hills to the south east of Edinburgh. A new 3.5-mile (5.6-km) pipeline under the local loch carries the water directly from the spring to the high-tech bottling plant on the farm owned by Stella Findlay, who, with her husband Marcus, is a principal shareholder in the company. Findlays was first marketed as a sparkling water, the carbonation being added during the bottling process. A still version was introduced a year later in response to customer demand. Initially, the main market for Findlays was selected catering and retail outlets in southern Scotland, but the company is developing a distribution network for its water throughout Britain.

Fionnar – the Gaelic name means 'fresh' or 'cool' – is a natural mineral water from the Fionnar Spring which rises high in the hills to the west of Loch Ness in Scotland. Exceptionally pure, it contains no nitrates and is very low in sodium. Supplied to North Sea oil rigs and other selected outlets in Inverness-shire and neighbouring counties, Fionnar is also among the top four UK bottled water export brands. The main importers are France, Belgium and Germany, where the water is much appreciated as an accompaniment to Scotch whisky.

Hadham, with sales in 1993 of 3 million litres (nearly 0.8 million US gallons) is among the UK's top 10 spring water brands. It originates 150 metres (nearly 500 feet), below the chalk layers of Hertfordshire's Ash valley, where there has been a source of water since Roman times. Low in salts and nitrates, Hadham is a slightly hard water with a fine, clean taste. Available in still and carbonated varieties, more than 10 per cent of its sales are door-to-door.

Hambledon is a spring water from the Hampshire village of the same name, which is known as 'the cradle of cricket' because the modern laws of the game were devised there. The water lies more than 100 metres below the chalk of the Hampshire Downs, and is bottled and carbonated at source. For more than a century, it was (and still is) used by Hartridge, the local brewer, in the manufacture of its soft drinks. In 1982, the company decided to market the water, and now sells around 1 million litres (0.264 million US gallons) a year – 30 per cent of it door-to-door.

Hazeley Down, which gained natural mineral water status in 1991, owes its purity and its pleasant, tangy flavour to the layers of chalk which protect its source and provide natural filtration. The water is drawn to the surface through boreholes 142 metres (464 feet) deep, and is bottled at source in a purpose-built plant at Hazeley Down, near Twyford in Hampshire. Towards the end of 1993, Hazeley Down was repackaged to appeal specifically to the catering and off-licence trade, and is served in such restaurants as London's L'Escargot and Café St Pierre. From being predominantly Hampshire-based, distribution now covers 50 per cent of Britain.

464, a sister brand, was launched in 1994. Also a natural mineral water, it is similar in composition to Hazeley Down, though drawn from another borehole of the same depth (hence the name). Sold through retail grocery outlets, it is designed for young adults with an interest in health and fitness. 464 made its debut as a still water in distinctive octagonal bottles with eye-catching red and blue labels. A sparkling variety is also planned.

Hildon is another British water that sells well overseas. In 1993, one third of its production was exported to Germany, Switzerland and Austria, but it has a firm following in its native Hampshire and throughout southern England. Customers at Terence Conran's famous restaurants such as Le Pont de La Tour and Quaglino's, find it is the only water on offer (unlike the huge choice of wines). Hildon is a natural mineral water with a source 100 metres (327 feet) beneath the chalk hills which border the River Test valley. A new bottling plant was opened in 1989 by the Duchess of York, and produces still and sparkling Hildon water in a variety of bottle sizes.

Isis, with its distinctive black and white label, was inspired by a goddess and discovered by a local water diviner. From an aquifer below the chalk and flint of the North Dorset Downs, Isis is brought to the surface through a well 50 metres (164 feet) deep, and bottled at source. Low in sodium and other minerals, it achieved full mineral water status in 1990, and is sold in still and sparkling versions mainly to the hotel and restaurant trade in Dorset and the south of England.

Prysg is the premium Welsh water you are increasingly likely to find in leading hotels and restaurants, not only in Wales, where it is the market leader in this sector, but also in England, the US and Japan. London's Savoy, Dorchester and Ritz hotels all pour Prysg, the first brand to gain natural mineral water status in Wales. A quarter of total sales are in still in the principality, where Prysg has become the official mineral water for all Welsh national rugby union squads. The water flows from the spring at a constant 8,400 litres (nearly 2,200 US gallons) per hour, and is bottled at source at Maesycrugiau, Pencader, south-west Wales.

Rydon Springs water from the village of Kingsteignton, south Devon, has, according to ancient records, provided refreshment for Roman legionaries, Anglo Saxon villagers, Norman knights and mediaeval monks. Its source, an artesian spring under the limestone and sandstone layers of the Haldon Hills, produces 12.73 million litres (3.36 million US gallons) a day of pure spring water. Rydon Springs was launched in 1992, and quickly built up its production level to 3 million litres (nearly 0.8 million US gallons) a year. The water is extracted by a gravity-fed 5.48-metre (18-feet) shallow bore, and carried by pipeline to the bottling plant. The still and sparkling water is sold in in south west England, London and the Midlands.

Somerley water comes from a spring which has served the 7,000-acre Somerley Estate, family home of the Earl and Countess of Normanton, for more than 200 years. The water, which is low in minerals and virtually sodium free, has been bottled on the estate in a modern, purpose-built plant since November 1992. The Somerley range includes still and sparkling versions of the spring water, plus a vitamin-enriched drink and five fruit-flavoured sparkling waters which are sold nationally through a wide range of outlets.

Tom Cobley's Dartmoor Spring Water from Dartmoor National Park may have refreshed the original Uncle Tom Cobley on his way to Widecombe Fair. It rises from an ancient source which was recorded in the Domesday Book. With an almost perfect acid/alkaline balance, the water has a crisp, clean taste and a low sodium and mineral content. Launched in May 1992, it is bottled at source in a high-tech plant with a capacity of 3,600 litres per hour. As well as the original still and sparkling varieties, there are now six fruit-flavoured waters which are expected to account for half the plant's total production in 1994.

Ty Nant is probably noticed more for its beautiful blue bottle than for its delightful flavour, though a panel of eight tasters recently voted it the overall best buy among 61 still and carbonated waters. This Welsh spring water from the hamlet of Bethania in the Cambrian Mountains was declared by the local water diviner, who discovered it back in 1976, to be the purest that he had ever tasted. Since the launch of Ty Nant in 1989 (and the British Glass award for the excellent design of its bottle), sales have increased to around 2.5 million litres (0.66 million US gallons) per year, 80 per cent of which is exported to the US, Europe, the Far East and South Africa. It is available in Harrods, the Café Royal, and other high-class stores and restaurants in the south-east of England.

Wight Crystal is a spring water from the Isle of Wight, known to generations of seafarers for its 'good keeping' water. Drawn from deep artesian wells under the island's central ridge of chalk downlands, Wight Crystal is low in minerals but has a distinctive, slightly peppery taste which is claimed to be due to its delicate balance of natural trace elements. Bottled at source in still and sparkling versions by Osel Enterprises Ltd, a company which provides work for people with disabilities, it is stocked by retailers, hotels and restaurants on the island and in southern England.

Ballygowan: still, moderately and highly carbonated, low mineralization.

BALLYGOWAN

This clean-tasting, hard water from County Limerick outsells all of its nearest rivals by a factor of three. With 70 per cent of the market, Ballygowan is far and away the brand leader among bottled waters in Ireland. Its popularity is not confined to its country of origin. In 1993 it accounted for more than 9 per cent of total imports of bottled water to the United Kingdom, where it features among the 15 best sellers, and it also has considerable sales in the US, Italy, Germany, Australia and Japan. "I'll have a Ballygowan," is now heard in Irish pubs and clubs around the world, especially those frequented by young people.

Ballygowan Spring Water Ltd was founded in 1981 by the company's present chairman, Geoff Read, but the source, located in the fertile farmlands of south-west Ireland near the small town of Newcastle West, was discovered more than 800 years ago by the Knights Templar, who named the spring St David's Well, and built a castle there. Subsequently the castle and lands passed to the Earls of Desmond, then to members of the Courtenay family through whom they descended to the Earl of Devon, until, in 1986 they became the property of the bottling company.

Drawn from an aquifer that is almost exclusively limestone, Ballygowan has relatively high concentrations of calcium, magnesium and bicarbonate, but is virtually sodium free. The aquifer is overlain with a thick layer of impermeable clay which protects the purity of the water, and the source and catchment area are free of industrial or agricultural pollutants.

Ballygowan is one of the four Irish brands which has official mineral water status, gained in 1988, although it is now marketed as a spring water, and has third place among the top selling spring water brands in the UK. The water is bottled at source in a fully automated, custom-built plant which is one of the most modern in the British Isles. The bottling line can fill up to 600 bottles a minute. Three versions of Ballygowan are produced – still, lightly sparkling and sparkling. The sparkling varieties have 5.6 and 7.5 grams per litre of CO_2 added during bottling.

Production: 30 million litres (7.9 million US gallons).

Exports: to the UK, Italy, Germany, the United States, Australia and Japan.

Bottling company:
Ballygowan Spring Water Co. Ltd,
Castle Demesne,
Newcastle West,
Co. Limerick,
Ireland.
Owned: Allied-Lyons plc and Guinness plc.

Analysis:	*mg/l*
Calcium	*114.0*
Magnesium	*16.0*
Sodium	*15.0*
Potassium	*3.0*
Bicarbonate	*400.0*
Chloride	*28.0*
Sulphate	*15.0*
Nitrate	*1.0*

GLENPATRICK SPRING

'Glenpatrick Spring Water Company pitches its product quite deliberately at the upper end of the market,' says managing director Brian Duffy. 'Our mineral water is clearly identified with all that is regarded as best in Ireland – fresh flowing streams, pure air, rolling green countryside and an unpolluted environment which have made Irish blood-stock and Irish agricultural products renowned throughout the world.'

Glenpatrick Spring aspires to this Irish tradition for quality. It is one of the few mineral waters to have won, in 1989, the supreme gold medal for excellence awarded biannually by the British Bottlers' Institute. The water, which is low in sodium and nitrates, has a pleasant, invigorating flavour which mixes well with wines, spirits and fruit juices or is refreshing drunk on its own.

Its source is located within the company's two-acre greenfield site on the outskirts of Clonmel, a small town in southern Tipperary, near a place of pilgrimage known as St Patrick's Well, where for centuries a clear natural spring has bubbled to the surface, disgorging 500 gallons of water an hour into an adjacent pool. The site is still marked by a rough stone cross and a ruined church. Glenpatrick Spring is pumped from an aquifer 46 metres (150 feet) below the surface of the Golden Vale, rich farming land under which there are extensive veins of limestone. The water is named after the neighbouring hamlet of Glenpatrick situated high in the Comeragh mountains which lie to the south of Clonmel.

After being brought to the surface, the water is bottled at source under strictly controlled conditions in the company's modern factory, to which has recently been added new laboratories and storage facilities. There is a still and a sparkling version of the water. The latter has the equivalent of 3.5 grams per litre of carbon dioxide added at the bottling stage.

Glenpatrick Spring was launched on the Irish market in April 1986. Three months later it made its appearance in New York, most notably in Macy's, Shoprite and Sloans. In September that year it arrived in the United Kingdom – the first Irish spring water to break into this market. It is now well established in affluent niche markets with discerning, health-conscious customers. In Britain, Glenpatrick Spring is stocked not only by Harrods and Selfridges, but by Marks & Spencer, under whose St Michael label a significant part of the company's production is sold both as a natural water and as a sparkling spring water drink with various fruit and other flavours, such as elderflower, added. Glenpatrick's success is really founded on such exports to many world markets, where, as with Vichy Celestins from France, people enjoy a little of the flavour of the country of origin.

Production: more than 4 million litres (1.057 million US gallons).

Exports: approximately 75 per cent of production mainly to the US, Canada and the UK.

Glenpatrick Spring: still and moderately carbonated, low mineralization.

Bottling company:
Glenpatrick Spring Water Company Ltd,
Cashel Road,
Clonmel,
Co. Tipperary,
Ireland.
Owned: Queally/Dawn Foods.

Analysis:	*mg/l*
Calcium	*112.00*
Magnesium	*15.00*
Sodium	*12.20*
Potassium	*1.10*
Bicarbonates	*400.00*
Chloride	*20.00*
Sulphate	*19.00*

Tipperary: still and highly carbonated, low mineralization.

TIPPERARY

Tipperary natural mineral water is another Irish recipient of the British Bottlers' Institute gold medal for excellence. Microbiologically, the water is remarkably pure. Apart from its high potassium content, which may contribute to its subtly refreshing flavour, Tipperary is low in minerals. It contains little sodium and calcium and no nitrates, and is ideal for mixing baby formulas, or serving to infants and invalids.

The source is at Borrisoleigh, which nestles between the Devil's Bit Mountain and the Rock of Cashel in the centre of County Tipperary – an area of lush, green pastures and wooded, gently sloping hills, free from environmental pollution. The water rises to the surface at a constant temperature of 11°C (51.8°F) from a single borehole which penetrates 100 metres (327 feet) through sloping layers of impermeable limestone to an aquifer in the Devonian sandstone beneath. It takes at least 40 years for the rainwater to percolate through the rock strata to the aquifer below, and the exceptional purity of Tipperary water is largely due to this lengthy filtration period.

The water is bottled and sealed directly over the source in a high-technology production plant which recently underwent an IR£2.5 million modernization programme. As many as 26,000 bottles per hour can be filled by the bottling unit, which delivers either still or sparkling versions of the water in recyclable PET and glass bottles varying in size from 200 ml to 2 litres. Nineteen-litre water cooler units and containers of Tipperary Mist, an ozone-friendly facial spray, are also produced in the factory.

The brand was launched in 1986 by the Tipperary Natural Mineral Water Company, a subsidiary of the Gleeson Group, a private company owned by Patrick and Nicholas Cooney which manufacturers and distributes soft drinks for the Irish home and export markets. A year later Tipperary gained natural mineral water status under European regulations, the first Irish bottled water to achieve this. It has since become a major national brand in Ireland, sharing nearly 90 per cent of the market with Ballygowan, Aquaporte and Glenpatrick, and has developed a number of export markets, among which are Russia and Japan.
Production: 1 million litres (0.26 million gallons).
Exports: 10 per cent of annual production.

Bottling company:
Tipperary Natural Mineral Water Company,
Pallas,
Borrisoleigh,
Co. Tipperary,
Ireland.
Owned: Gleeson Group.

Analysis	*mg/l*
Calcium	*37.0*
Magnesium	*23.0*
Potassium	*17.0*
Sodium	*25.0*
Bicarbonates	*282.0*
Sulphate	*10.0*

ICELAND
INTRODUCTION

For an avant-garde country, Iceland has come late to the bottled water business, probably because of its distance from the rest of Europe – it is 180 miles from its nearest neighbour, Greenland. Also, with only 260,000 inhabitants, its home market is very small, and its tap water so good that Icelanders hardly feel the need to buy a product that is so abundantly there for free. But in recent years, increasing travel and tourism have made them much more aware of the eating and drinking habits of other countries, especially the thirst for bottled water that is apparently unquenchable.

Iceland's comparative remoteness, its lack of intensive agriculture and heavy industry, and its system of winds and tides which protect it from acid rain and other weather-borne pollutants, make it the cleanest country in the world. As Icelanders are beginning to recognise, it is an ideal setting from which to market its exceptionally pure water to an environmentally conscious world. There are now four water bottling companies in Iceland, three near the capital, Reykjavík, and one in the north, at Akureyri. Probably the two best known brands, especially in Britain and the US, are Svali and Thorspring. Both are spring waters. Svali comes from the Bishop Gwendur Springs near Reykjavík, and is bottled by Islenskt Bergvatn, and Thorspring, from another source in the same area.

THORSPRING

Thorspring is 99.993 per cent pure, and claims to have one of the lowest levels of undesirable chemicals ever recorded. It is an extremely palatable, alkaline still water, with a very low mineral content, and is sodium- and nitrate-free. The source lies in Heidmörk, a protected nature reserve, on the outskirts of Iceland's capital, Reykjavík. The area is one of small streams which now and then surface as springs. A 400-metre (1,300-feet) deep borehole has been drilled upstream of Thorspring's natural spring, through the volcanic rock strata, to the aquifer below. From the well-head the water is piped 8 kilometres (5 miles) to the bottling plant in Reykjavík. The bottling company is owned by Hof Holdings (Iceland's largest retailer), Vifilfell Ltd (the bottler for Coca Cola in Iceland), both of which have a 45.5 per cent stake, and by the Reykjavík Municipal Waterworks, which holds a 9 per cent share.

Launched in 1990, Thorspring has been bottled and marketed for nearly four years.

"It is intended only for export, since the water in Iceland is so pure that we can drink it from the tap. No water is sold in the stores here except for a carbonated one," Dröfn Thorisdóttir, the company's general manager, told us. So far exports have been only to the United States – 3 million litres (0.8 million US gallons) were sold there, mainly in the mid west, in 1993, a 150 per cent increase on the previous year. Production: 3 million litres (0.8 million US gallons).

Thorspring: still, low mineralization.

Bottling company:
Thorspring-Iceland Inc.,
Studlahals 1,
110 Reykjavík,
Iceland.
Owned: Hof Holdings, Vifilfell Ltd and the City of Reykjvik.

Analysis:	*mg/l*
Calcium	6.2
Magnesium	0.5
Sodium	8.3
Potassium	0.4
Chloride	11.0
Sulphate	2.5
Bicarbonate	28.0

THE NETHERLANDS

INTRODUCTION

Unlike its nearest neighbours, Belgium and Germany, The Netherlands has no continuing history of spa culture or mineral water appreciation. In fact, the Dutch mineral water market has almost been an extension of the Belgian. Many of the great Belgian producers, like Spadel and Chaudfontaine, label their waters in both French and Dutch for sale in either country. Of the 224 million litres (59.45 million US gallons) of bottled water drunk in Holland each year, almost three-quarters is imported, mostly from Belgium, though small amounts also dribble in from France, Germany and countries further afield. Belgium's Spa is far and away the market leader in Holland, and has been supplied under Royal Warrant to the Dutch as well as the Belgian court for many years.

Development of the Dutch market has been hampered to some extent by the unfortunate decision of the The Netherlands government in 1985, to deal with both natural mineral and spring waters in the same regulation, with the result that some of the requirements applicable only to mineral waters have become mixed up with those for spring waters, and vice versa. This should be sorted out when the pending amendment to the European directive is converted into Dutch law.

However, The Netherlands has a number of good home-produced waters, which are beginning to gain greater appreciation within their country of origin. Seven mineral and eight spring water sources are now officially recognized. Among the more popular waters currently available in Dutch supermarkets and restaurants are Bar-le-Duc and Cathareine, both from the Raak stable, Sourcy from Vrumona, Sablon from Dranken Industrie Sittard (DIS), which also produces Sourcy under licence, Maresca from Frisdranken Industrie Winters, and Hébron, produced by Hero Nederland, which is also sole agent for the sale of France's Perrier and Volvic in Holland. The first five are mineral waters officially recognized under the terms of European directive 777, and Hébron is a spring water which gained recognition as such in 1990.

The Dutch market has grown enormously over the past 20 years from very small beginnings. In 1975 the annual per capita consumption of bottled water in Holland was just over a litre. Although it has fallen off a little from the the 15.6 peak of 1989, it still hovers around the 15-litre mark – half as much again as the British manage to imbibe. The current split is 75:25 in favour of the sparkling brands. Over the past few years the Dutch have been downing proportionately less imported water, and exporting a smaller percentage of the home-produced variety, which means that they are consuming greater quantities of their own bottled waters. The move in Holland, too, is away from spring towards premium brand mineral waters in a reversal of the recent UK trend.

National Association:
Vereniging Nederlandse Frisdranken Industrie (NFI),
Heemraadssingel 167,
30-22 CG Rotterdam,
The Netherlands.

BAR-LE-DUC

Bar-le-Duc, which has been marketed in The Netherlands since 1906, is a fairly lightly mineralized water. Low in sodium, it has useful amounts of calcium, chloride and sulphate. The water is pumped through a borehole from a depth of 190 metres (620 feet) from one of the Bar-le-Duc springs which are situated on the Dutch /Belgian border in the village of Baarle-Nassau/Baarle Hertog. The village, which lies in an area of heath and woodland, is unique in Europe because of its enclaves - small pieces of Belgian land completed surrounded by Dutch territory. The bottling plant is, in fact, partly in Holland and partly in Belgium. The other Bar-le-Duc source produces spring water. Both were officially recognised by the Dutch Ministry of Health in 1987, and between them produce some 60 million litres (15.85 million US gallons) a year, about 60 per cent of which is mineral water. The latter is available in still and sparkling varieties, the carbonated version having 6 grams per litre of CO_2 added at the bottling stage. The sparkling version is sold in 1-litre glass bottles, and the still water in 2-litre cartons.
Production: 36 million litres (9.5 million US gallons).

Bottling company:
Raak (Holland) BV,
Reactorweg 69,
3542 AD Utrecht,
The Netherlands.

CATHAREINE

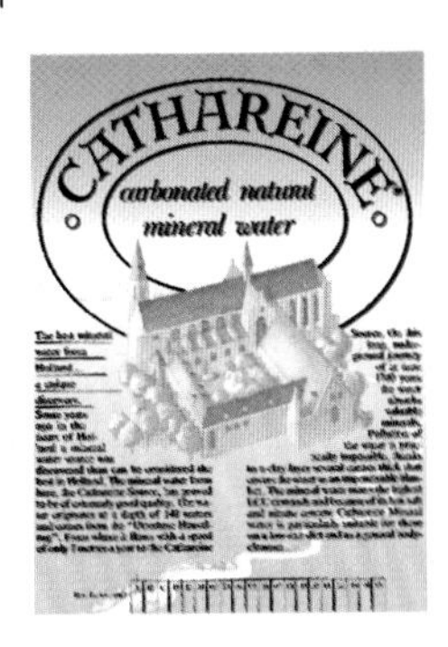

Cathareine is the sister brand to Raak's Bar-le-Duc. Another lightly mineralized water, its source is located at Utrecht in the centre of the country. Water from the Utrechtse Heuvelrug, an underground reservoir, flows at a rate of 7 metres (23 feet) a year into the Cathareine source, 43 metres (140 feet) below the surface. On this long, underground journey, which takes at least 1,700 years, the water absorbs valuable minerals. A layer of clay, several metres thick, protects the source from possible contaminants. Cathareine, which gained official mineral water status in 1986, is bottled at source. Still and artificially carbonated versions of the water are available. Its low sodium and nitrate content makes it particularly suitable as a general body cleanser and for those on low-salt diets.
Production: not available.

Bottling company:
Raak (Holland)BV,
Reactorveg 69,
3542 AD Utrecht,
The Netherlands.

HEBRON

Officially recognized as a spring water source in 1990, Hebron has since been sold in supermarkets throughout The Netherlands. It is a slightly sparkling, lightly mineralized water, with 385 gm/l total dissolved solids, contains virtually no sodium or nitrates, and has a healthy dash of calcium (97 mg/l) and magnesium (17 mg/l), plus a veritable cocktail of interesting trace elements. Hebron is drawn from Hero's Hoensbroek source, and complements the company's two imported French mineral waters, Volvic and Perrier.
Production: 9 million litres (2.38 million US gallons).

Bottling company:
Shiffers Food BV,
Wijngaardseweg 59,
6433 KA Hoensbroek,
The Netherlands.
Owned: Hero Nederland BV.

Ramlosa: moderately carbonated, medium mineralization.

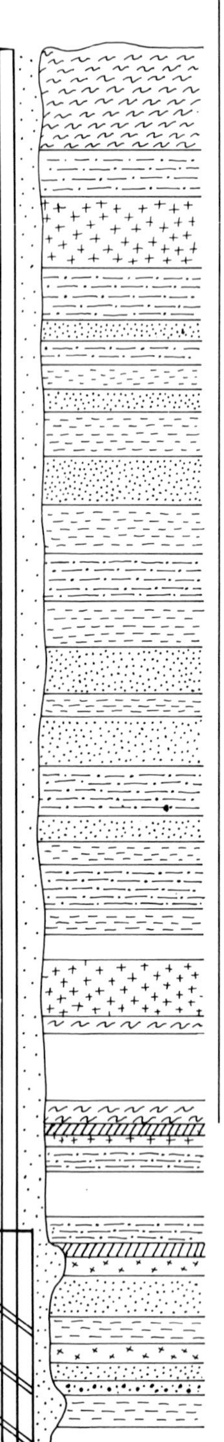

SWEDEN: RAMLÖSA

Ramlösa projects both a royal and a sporting image. This alkaline water from the south west of Sweden, just by the seaport of Halsingborg, carries the official warrant of purveyor to HM the King of Sweden and HM the Queen of Denmark. As the natural choice for athletes, Ramlösa was appointed official mineral water of the Olympic Winter Games in Lillehammer in 1994. Ramlösa is part of the Volvo BCP Group, which is owned by AB Volvo.

In Sweden Ramlosa and mineral water are almost synonymous; Ramlösa has 95 per cent of the market for natural mineral waters. It is, however, a small market. Although the Swedes pay exceptionally high prices for alcohol, and face some of the world's toughest drinking and driving laws, mineral water consumption is a mere 11 litres (2.9 US gallons) annually, far below the European norm.

Ramlösa, however, has survived as a therapeutic mineral water which dates back nearly 300 years. The founder of the Ramlösa Spa was a physician named Johan Jacob Dobelius, who noticed that local people attributed therapeutic qualities to a local spring. He examined it, and was sure he had found a genuine, natural mineral water. On 17 June 1707, to honour the birthday of Sweden's King Charles XII, he took a party to sample the waters. Thereafter, the reputation of the spring grew. People from all walks of life came to

Key:

Sandstone, medium & coarse
Sandstone, fine
Siltstone, coarse
Siltstone, fine
Mudstone, with silt
Mudstone, interlayered clay
Claystone, shale
Clay inclusion
Carbonaceous shale, coal seam.

Johan Jacob Döbelius, who discovered the Ramlösa mineral water spring.

'take the waters'. The appeal of the waters made Ramlösa the most fashionable spa in northern Europe during the 18th and 19th centuries.

But times changed, and as the government extended its control over health services for all Swedes, it became necessary to close down the activities of the Ramlösa spa in 1973. The bottling plant, which had been situated in the spa area, was also moved. A new bottling plant was built nearby, and today, the water is pumped via a stainless steel pipe from the underground spring in the spa area to this modern bottling unit. Production is under the supervision of the local Environment and Health Office, authorized by the Swedish Food Administration. During the 1980s production quintupled, and has reached more than 60 million litres a year.

Geological studies have revealed that the Ramlösa aquifer is in the water-bearing sandstone layers at a depth of over 90 metres (295 feet), protected by a diversity of more than 30 overlaying beds of sand, siltstone, mudstone, clay and even a coal seam. This multi-layered bedrock, which was formed about 180 million years ago, seals off the aquifer, so that is is recharged with water very slowly. One geological study suggests that the water in the aquifer may be 70 to 80 years old.

Until the beginning of this century, Ramlösa was distributed as a still water (as it emerges), but it is now moderately carbonated with 5 grams per litre. It is marketed overseas in the traditional distinctive blue glass bottles, which are also used for a small proportion of home market sales. In Sweden, Ramlösa is sold mainly in returnable clear glass and PET bottles.

Production: More than 60 million litres (16 million US gallons).

Exports: 40 per cent of production to 35 countries.

Bottling company:
AB Ramlosa
Hälsobrunn,
Box 15023,
S-250 15 Helsingborg,
Sweden.
Owned: AB Volvo.

Analysis:	*mg/l*
Chloride	23.0
Fluoride	2.8
Sulphates	7.3
Potassium	1.5
Sodium	222.0

Vichy Original: highly carbonated, high mineralization.

Vichy Novelle: moderately carbonated, medium mineralization.

Victor: highly carbonated, high mineralization.

Bottling company:
Hartwall-Juomat Oy,
Konalantie 47 B, PL 23,
00391 Helsinki,
Finland.

Vichy Original:	*mg/l*
Potassium	*240.0*
Magnesium	*110.0*
Calcium	*100.0*
Sodium	*220.0*
Bicarbonate	*800.0*
Chloride	*480.0*
Sulphate	*160.0*

Vichy Novelle:	*mg/l*
Potassium	*120.0*
Sodium	*1.0*
Magnesium	*110.0*
Calcium	*70.0*
Chloride	*120.0*
Sulphate	*450.0*
Hydrogen carbonate	*150.0*

FINLAND: HARTWALL

Finland's most popular mineral water, Vichy Original, is not, in the European sense, a mineral water at all. It is neither extracted from nor bottled at a recognised source. It is a 'manufactured' water, produced, like beer and soft drinks, with a number of carefully measured ingredients. Nevertheless, in Finland's shrinking mineral water market, Vichy Original and its two stable mates Vichy Novelle and Victor, account for more than a third of the country's 30 million-litre (7.93 million US gallons) annual consumption of bottled water.

The mineral content of Vichy Original, the brand leader, bears some resemblance to Vichy Celestins, from which it probably took its name. It contains a fair amount of sodium and bicarbonates, though not nearly so much as its French namesake, and a good balance of other minerals. It is highly carbonated with 8.6 grams of carbon dioxide per litre. It is a good accompaniment to food, and an excellent thirst quencher. Vichy Novelle, a relative newcomer to the market, was launched in 1990 to appeal to younger, health-conscious consumers. With 5 grams per litre CO_2, it has less fizz, and is virtually sodium-free, but its sulphate level is nearly three times that of Vichy Original. Victor, with less salt, but about the same carbonation level as Vichy Original, is regarded in Finland as having an extra fine taste, and is often drunk at meals instead of wine or beer.

The three waters are produced by Oy Hartwall Ab, Finland's oldest mineral water company, and one that for some time has also had interests in the brewing and soft drinks industries. The firm was founded in 1836 by scientist and businessman Victor Hartwall, who, with his former university tutor, mineralogist Professor P A von Bonsdorff, pioneered the commercial exploitation of artificial mineral waters in Finland. The Hartwall family has been involved in the company for five generations.

Why produce artificial waters in a country so well endowed with natural sources? The reasons were basically economic. Many of Finland's spas and sources are remote and were difficult to reach, especially in the early 19th century; the spa season was short and coincided with the busiest time in the rural calendar; spa waters were often needed at other times of the year; and imported waters were expensive and their availability and quality uncertain. So the practical Finns began to produce mineral waters artificially in centres of population like Helsinki and Turku where demand was high and transport costs low. At first they copied the contents of the famous French, German and Belgian sources, and offered them for sale as such, making it clear that they were artificially produced. In fact, the artificiality was a strong selling point. These waters became so popular that in 1838 an artificial mineral water spa was opened in Helsinki's Kaivopuisto park, and the demand in Finland for manufactured waters has remained.

Production: 10.5 million litres (2.8 million US gallons) for the three waters.

Radenska: naturally moderately carbonated, very high mineralization.

SLOVENIA: RADENSKA

Three red hearts are the symbol of this sharp, strong tasting, naturally carbonated water that outsells all others in Slovenia (the former north western republic of Yugoslavia), and is now widely exported. **Radenska** comes from the mineral springs at Radenci, which lies near the Austrian border. Medical student, Karel Henn, first noted their properties in 1833. Bottling started in 1869, in tall stone jars. The first visitors arrived to take thermal baths and cures in 1882, and at the turn of the century, Josef Hohn, an Austrian doctor, wrote a treatise on the diseases most susceptible to help from the waters.

Radenci is in the centre of a former volcanic area, where mineral water is found a mere 15 metres (50 feet) below ground. Carbon gas works its way up from the central magma and as it travels towards the earth's crust joins a large water layer; the resulting carbonic acid absorbs minerals from surrounding rock formations. Radenska is high in sodium, calcium, magnesium and sulphates, with the total mineral content around 3,500 mg/l.

The company owns 94 springs, 25 of which are in use for bottling mineral water, soft drinks and health waters, for extracting carbon gas for the company's own purposes and for sale to other bottlers, and for providing hydrotherapy. Radenska runs a large spa complex, with water treatments for many complaints. It also serves as a popular holiday and health resort for people of all ages.

During the last few years, the company has started to produce **Miral** and **Radin**, waters with less carbonation and fewer minerals than the market leader, Radenska. Miral, a sparkling water contains 2.8 grams of carbon dioxide and and 2,422 milligrams of minerals per litre. Radin is a still water with less than half Miral's mineral content.

There are eight bottling lines with a capacity of 1.3 million glass and PET bottles a day. The bottling plant employs 500 workers, although the grand total for all the Radenska enterprises is 1,460 jobs.

Production: Radenska 83.6 million litres (22 million US gallons); Miral 2.6 million litres (0.7 million US gallons); Radin 4.5 million litres (1.2 million US gallons).

Exports: Radenska 26 per cent of production to 11 countries; Miral 58 per cent of production and Radin 80 per cent of production, both to Russia, Croatia and the Czech Republic.

Bottling company:
Radenska TRI SRCA,
Zdravilisko naselje 14,
69252 Radenci,
Slovenia.

Analysis:	*mg/l*
Sodium	*470.0*
Potassium	*95.0*
Calcium	*217.2*
Magnesium	*96.7*
Chlorides	*47.7*
Sulphates	*88.4*
Bicarbonates	*2,365.0*

POLAND
INTRODUCTION

Like Germany on its western flank and Russia to the east, Poland has a long history of spa culture and mineral water consumption. Mineral water springs are plentiful throughout the country from the broad Polish plain in the north to the Tatra Mountains in the south. From the 18th century, dozens of spa towns were established where the wealthy and fashionable bathed in or drank the beneficial waters, much as the English did at Buxton and Bath. Among these are Ciechocinek, still Poland's leading health resort, which the composer Chopin was said to have visited incognito in a vain attempt to cure his consumption. There is also, for example, Naleczow, near Lublin, to the south-east of Warsaw, which during the last century attracted visitors from all over eastern Europe to its hot spring, set in beautiful parklands; Krynica, known as the pearl of Polish spas from its location at the base of the Tatra Mountains in southern Poland; Busko, between Warsaw and Cracow, in the south east, renowned for its curative peat and sulphur treatments; and Polczyn, in the north west, near the Baltic coast.

Some of the sources are too highly mineralized for drinking, and are used for therapeutic purposes only. But many others, like Krystynka, though they contain relatively high amounts of minerals, can be drunk quite safely, and may be valued for their sodium or calcium content. In line with western European tastes, lighter waters are becoming more popular in Poland, and are attracting foreign venture capital to improve and expand their production facilities. Naleczowianka, the subject of a joint enterprise set up by Nestlé Sources International and the East Bridge group in March 1994, is one example of this trend.

KRYSTYNKA

Krystynka: still and lightly carbonated, high mineralization.

Bottling company:
Uzdrowisko
Ciechocinek,
ul. Kosciuszki 10,
87-720 Chiechocinek,
Poland.

Analysis:	*mg/l*
Sodium	*900.00*
Potassium	*14.00*
Calcium	*176.00*
Magnesium	*59.80*
Chloride	*1,641.50*
Sulphate	*38.00*
Hydrogen Carbonate	*456.60*

Krystynka is a natural mineral water from the spa town of Ciechocinek in central Poland. Situated in the picturesque valley of the river Vistula, 80 kilometres (50 miles) north west of Warsaw, Ciechocinek has for many years been Poland's leading health spa. It was well known as early as the 13th century for its numerous thermal springs, which brought relief to sufferers with fractures, painful joints, lung complaints and blood circulation disorders. By the end of the last century, Ciechocinek was attracting visitors from many countries who came to enjoy the resort's sheltered, sunny climate as well as to bathe in the therapeutic waters. A well equipped sanitorium was built in 1967 with a wide range of medical facilities. Many of Ciechocinek's springs are too highly mineralized for drinking. But early in the century, it was discovered that the water from one closely resembled Germany's Bad Kissinger Theresienquelle. The Polish water, now known as Krystynka, is an excellent thirst-quencher with a pleasant astringent taste. The first bottling plant was built in 1903, and there was considerable demand for the water.

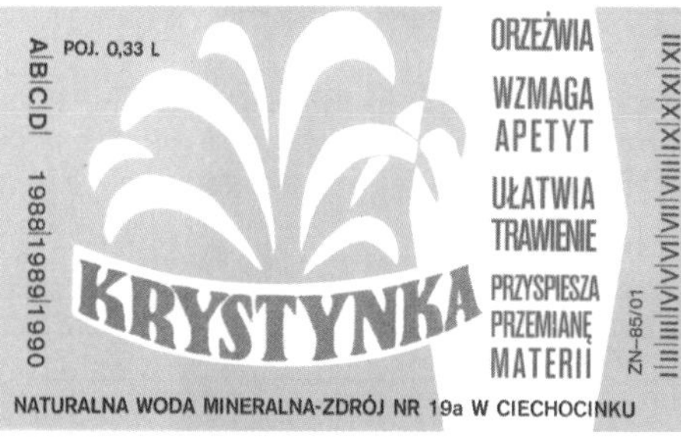

Subsequently, more sophisticated equipment was imported from Sweden, and arrangements were made for carbon dioxide, 2.85 grams per litre of which is added to the sparkling version of the water during the bottling process, to be transported from Kislovodsk in Russia. The purity and quality of the water are subject to regular monitoring by the state laboratory.

With its high sodium content, Krystynka is recommended as a drink for athletes and for those who work in hot climates or high temperatures. Available in 330-ml glass, and 1.5-litre PET bottles, it is also popular as a table water in Polish homes, hotels and restaurants.

The same company produces Kujawianka, which like Krystynka, is available in both still and sparkling versions. Kujawianka comes in 330-ml glass and 1.5-litre PET bottles. Both waters are at present sold only in the home market.
Production: 19 million litres (5 million US gallons) a year.

NALECZOWIANKA

Naleczowianka: naturally moderately carbonated, medium mineralization.

Naleczowianka's source is an ancient hot spring near the village of Naleczow, 140 kilometres (88 miles) south east of Warsaw. Located in the centre of a vast park containing a variety of rare trees, the spring has been famous since the early 19th century when it was fashionable for members of Polish and other European high society to gather there to sample the water. A calcium-magnesium water, which is low in sodium, Naleczowianka is naturally carbonated, and is appreciated for its stimulating, fresh taste. It is available in most regions of Poland.
Production: not available.
Bottling company:
Naleczowianka SP ZOO, Kolonia Botchotnica 5, 24-140 Naleczow, Poland. Owned: Nestlé Sources International and East Bridge group.

Analysis:	*mg/l*
Calcium	*119.3*
Magnesium	*24.3*
Sodium	*20.5*
Potassium	*5.0*
Hydrogen carbonate	*523.5*
Chloride	*12.1*

ROMANIA

As a result of its complex geology, Romania has an enormous number and variety of mineral waters springs which are found all over the country. Adrian Feru of the Romanian Administraton of Mineral Waters (RAMIN) likens subterranean Romania to 'a huge laboratory in which nature experimented to produce the ideal mineral water'.

The use of mineral waters for therapeutic purposes dates back at least to Roman times, and at present there are more than 100 spas and hydrotherapy resorts in the country. Medicinal and table waters have been bottled and distributed in Romania for more than 200 years. The first bottling plant was set up in Borsec in 1804, and by 1896, 20 small factories had been established. In 1989 annual consumption of mineral water in Romania had risen to 325 million litres (86 million US gallons).

Owing to its recent economic upheavals, Romania is one of the few European countries where demand for mineral water has fallen over the past few years – from 14 litres a head in the late 1980s to 6 litres in 1993. Production last year totalled 140 million litres (37 million US gallons). Now there are 25 bottling units in operation, owned by 17 different companies. These all bottle at source in accordance with European directive 777, and produce only carbonated waters, such as Borsec, Harghita and Perla, mainly for the Romanian market. The Romanian liking for *spritz*, explains Adrian Feru, is partly due to German influence and partly to the fact that Romania has so many naturally carbonated sources. However, like Britain, Italy and France, it seems that Romania is developing a taste for still, less mineralized waters. Two bottling plants under construction at Vatra Dornei and Domogled will produce these.

In 1990, the Romanian Administration of Mineral Waters (RAMIN), was set up to regulate, manage and market hydromineral resources owned by the state. There is also legislation which requires mineral water sources used for bottling and hydrotherapy to be surrounded by adequate protection areas.

Borsec is one of Romania's oldest and best-known mineral waters. It has been famous for well over 200 years for its pleasant, refreshing taste and for its beneficial effects on the digestive organs. Before a bottling plant was established at Borsec in 1804, wooden casks filled with the water were transported in covered, horse-drawn, wagons all over Romania. During the 19th century, the water won a number of international awards, and Emperor Franz Josef dubbed it 'the queen of mineral waters', from which the state bottling company has taken its name.

The water is drawn from a limestone and dolomite aquifer overlaid with clay, and surfaces through a natural spring and two drilled wells, at a height of 950 metres (3,100 feet) in a densely wooded area of the Carpathian Mountains, far from any potential pollutants. Borsec, which is naturally carbonated, is bottled only in glass with added

carbon gas from the source.
Production: 25 million litres (66 million US gallons).

In less than 20 years, the waters of Sincraieni have become well-known throughout Romania for their pleasant taste and balanced mineral content, the result of their long percolation through deep fissures in the volcanic and carboniferous rocks which characterize that area of the Eastern Carpathians.

Harghita, a naturally carbonated water, much appreciated for its uniquely piquant flavour and its perfectly balanced mineralization, is bottled in elegant turquoise green glass bottles with between 2 and 4 mg per litre carbon gas. The modern bottling plant's capacity is currently 30 million litres (7.9 million US gallons) a year.
Bottling company: Kraiten in Harghita JS, Str. Principala nr. 517, Sincraieni, Judetul Harghita, Romania.

Perla was probably drunk by 2,000 years ago by invading Romans as the source is very close to the ruins of a former Roman camp. Besides its excellent qualities as a table water, Perla is also recognized in Romania as a *Heilwasser* (health water) and recommended for digestive and other disorders. Naturally carbonated, it is bottled at source in a new high-tech plant with a bottling capacity of 15 million litres (3.96 million US gallons) a year. Perla is available in a regular sparkling version and as a semi-still water.
Bottling company: Vitarom Impex Ltd, Str. Garii nr. 600, Sincraieni, Judetul Harghita, Romania.

Apart from these, there are a number of other Romanian mineral waters, drunk either for refreshment or as medication. All are naturally carbonated with between 1.5 and 3.5 grams per litre of carbon dioxide.

Dorna is a calcium-rich water from the Poiana Negri source, bottled in glass by Apemin SA at Vatra Dornei.

Biborteni, a highly mineralized, calcium-magnesium water, has been on the market since 1900. Famous throughout the country for its pleasant flavour and curative properties, the water is bottled at source by Bibco SA. At present available only in glass, Biborteni will also be bottled in PET from 1995.

Buzias is another water much appreciated for its effectiveness in dealing with digestive upsets and other complaints. It comes from the spring of the same name, and is bottled by Phoenix SA.

Imperiala was discovered in 1789, the year of the French Revolution. Referred to as Aqua Imperatorum (the Water of Emperors), it comes from the Baciia source in the Transylvanian Alps, and contains a rich mix of minerals among which calcium, sodium, chloride and sulphate predominate. Imperiala is bottled by UIPSAP.

National Association:
Romanian Administration of Mineral Waters (RAMIN),
Mendeleev Str. 34-36,
Bucharest, Romania.

GREECE

When we first visited Greece in the 1960s most small restaurants invariably placed a carafe of water on the table along with the wine. If you hesitated and wondered about bottled waters, the waiter always responded that Greece had wonderful water, there was no need for bottled mineral waters. Indeed, they took it as an insult.

Thirty years on, however, the vogue is catching on fast. The market for bottled waters was estimated at 272 million litres (71.5 million US gallons) in 1993 (up an astonishing 21 per cent on the previous year) which is 26 litres per capita; not quite up to Italian or German levels, but far ahead of the UK or the Netherlands, and in hot pursuit of the United States and Portugal.

The best-selling Greek water is **Loutraki**, owned by Pepsi-Cola no less. It is in a league of its own with 35 per cent of all sales. Loutraki comes from an ancient spa town, once know as Therma, at the east end of the Bay of Corinth. Its famous, highly saline, healing waters, gush from the mountainside, close to the sea, and emerge at a warm 30°C (86°F). Ancient Greek writers praised these waters for their effectiveness against various ailments, including arthritis and liver complaints.

By contrast, the Loutraki spring, one of the oldest in the world, has a very low sodium content. A fresh tasting, hard water, mentioned in Roman and Byzantine as well as Greek literature, it surfaces at 25°C (77°F), and is moderately mineral-ized with a fair helping of magnesium, bicarbonates and silicates, gathered during its percolation through layers of limestone and gravel. Annual sales throughout Greece have now reached around 80 million litres (20.6 million US gallons).

Serious competition is now in prospect from **Korpi**, which has been bottled for some years, but was bought by Nestlé Sources International in 1993. Low in sodium and nitrates, this moderately mineralized water has been famous locally since the beginning of the century for its digestive and diuretic properties, and is now the second most popular brand in Greece.

The cold water source surfaces at a height of 260 metres (850 feet), near Vonitsa in the mountains of the northern Peloponnese. Korpi is bottled in still and carbonated versions and distributed through-out the country. Production is around 40 million litres (10.6 million US gallons).

Hayat: still, low mineralization.

TURKEY: HAYAT

Hayat, Turkey's leading brand of mineral water, is also sold in Japan under the name of Umashi Mizu. This exceptionally pure water, with a very low sodium and mineral content, comes from the Sekerpinar spring which rises at a height of 950 metres (just over 3,000 feet) in the Taurus mountains of southern Turkey. The spring is fed by glaciers at the summit of the mountain range, and is filtered underground through marble rocks. The mountain peaks, which rise to a height of 3,000 metres (nearly 10,000 feet) above the spring, are totally uninhabited with neither farms nor livestock, hence the purity of the water.

The source has been famous since pre-Christian times, and is said to have been visited in the third century BC by Alexander the Great, probably during his first campaign against the Persian King Darius III, and some 250 years later by Egypt's Queen Cleopatra, when she was summoned to the region by the Roman general, Mark Anthony.

Hayat Spring Water is bottled at source by Marsa Kraft General Foods, a joint venture between Jacobs Suchard of Kraft Corporation and Marsa Inc. of the Sabanci group of companies, one of the largest conglomerates in Turkey. The bottling plant is located at Pozanti, Adana, in the foothills of the Taurus Mountains, about 32 km (20 miles) from Tarsus, the birthplace of St Paul.

The plant employs the latest modern technology in its bottling system. All machinery is made from stainless steel, and the filter systems are imported from Switzerland. The internal pressure of the plant is higher than the outside atmospheric pressure to prevent any contamination of the sterilized bottling areas, and nothing is added to the water during the bottling process.

For the Turkish home market, in which Hayat has a 40 per cent share, the water is available in five sizes of PET bottle, ranging from 5-litres to 0.33-litre. It is exported to Singapore, Japan and other countries of the Far East (under the name of Umashi Mizu), to Europe and to the Turkish Republic of Northern Cyprus in 1.5-litre and 0.5-litre PET bottles.

Production: 95 million litres (25 million US gallons) a year, to be increased four-fold during 1994.

Bottling company:
Hayat Kaynak Suyu Fabrikasi,
Alpu Koyu Mevkii,
Pozanti Adana,
PK1, Turkey.
Owned: Marsa Kraft General Foods Sabanci Food Industry and Trade Inc.

Analysis:	*mg/l*
Calcium	*23.2*
Magnesium	*5.4*
Sodium	*3.7*
Potassium	*0.3*
Bicarbonates	*107.0*
Sulphates	*8.2*
Chlorides	*5.4*
Nitrates	*1.3*

UNITED STATES
INTRODUCTION

Berkeley Springs, West Virginia were made famous by George Washington.

The week after Bill Clinton won the US presidential election in November 1992, *People* magazine ran a story on what was 'out' of fashion and what would be 'in' with the new president. The saxophone, which he plays, was naturally 'in', so were sweet potatoes and Mountain Valley Spring Water from his home state of Arkansas which, *People* noted, he 'guzzles ... for his chronically hoarse voice', and to keep him going when he is out jogging. This put the presidential seal on what has been easily the fastest-growing drinking trend in the United States over the last decade. As the International Bottled Water Association put it in a press release headline for its 1993 convention in Fort Lauderdale, 'Bottled Water Increases Share of Stomach'. Wall Street analyst Sally Schaadt reported that bottled waters 'is the category to watch in the 1990s ... (it) is at the heart of the revolution in the beverage market. Consumer tastes have changed. Clean, clear and 100 per cent natural is the choice for the 90s.'

The evidence of this 'health wave', as one wag christened it, is that US consumption doubled from 4.6 gallons (17.4 litres) per capita in 1984 to around 9.5 gallons (36 litres) a decade later (by comparison consumption of coffee, alcoholic drinks and even milk all declined). By 1994 US consumption was more than 2.1 billion gallons (8 million litres) worth $3 billion wholesale. Ninety per cent of the sales are in still water.

The catalyst that really got America going, however, was a sparkling water – Perrier. The launching of a $6 million national advertising campaign by Perrier in 1977 stimulated unprecedented growth. Historians of the water boom divide the 'pre-Perrier years' from the 'high profile Perrier years'. Perrier quickly became a cult, spawning jokes like the one about the chap, eager to order the latest 'in' drink, who asked for 'a Perrier and soda'. The impact of the Perrier campaign, however, was not only its own sales. Its very success made the owners of many American and European waters realise that their bottles, too, could be advertised from coast to coast. 'Everyone climbed aboard the water-wagon Perrier set rolling,' recalled an American water bottler.

The timing was perfect. As health food stores came into fashion, so did natural waters (which had been quite forgotten since the days when Americans flocked to their own spas in the late 19th century). The vogue was encouraged by fears of pollution in, and the terrible taste of some municipal water supplies. Natural disasters, such as the earthquake in southern California in 1994, provide added impetus, not only making people nervous of local supplies, but even keeping more bottled water at home against such events (a replay of what has long been the custom in Italy of stocking up on bread, wine and water in any crisis). Today, bottled water is also being chosen as a drink in its own right, like soft drinks, coffee or fruit juices, and not just as an alternative to tap water.

Old poster lauds Claifornia's Bartlett Mineral Springs.

'The big boom,' said Lisa Prats of the International Bottled Water Association, 'is in single serving sizes (i.e. small bottles), not just in restaurants, corner delis or convenience stores, but in children's lunch boxes.' The trend was confirmed by Kim Jeffrey, chief executive of the Perrier Group of America, who told us, 'It's tremendous, tremendous. We had 80 per cent growth in 1993, and we'll get 80 per cent again in 1994. We've come from nothing to $100 million sales of small bottles, just for our group." The benefit (at least to water bottlers) is that the mark-up on small bottles is much greater than on the large plastic bottles and jugs, often of 5 gallons, for use in water coolers that are very much the American tradition. Gallon for gallon, these bulk sales in big polycarbonate or glass bottles filled with spring or processed waters still account for almost 70 per cent of all US water sales. 'Remember all those water coolers with paper cups you've seen movie stars like Humphrey Bogart sipping from in Hollywood movies?' John Mee of the Mountain Springs Water Company in San Francisco pointed out to us when we first investigated American waters in the mid-1980s. 'That's the real image of the bottled water business in America.' Although

Perrier, Évian and other imported European waters have gone some way to change that concept and have undoubtedly created a demand for 'premium waters', the daily business for many water bottlers is delivery in bulk to home and office. When we approached the Suntory Water Group, which ranks fifth in US bottled water sales, for information on their waters for this guide, they promptly dispatched us an impressive package about Crystal Springs, Belmont Springs and Polar Springs, but all illustrated in brochures for 3 gallon or 5 gallon bottles along with assorted coolers, dispensers, wicker bottles, mugs (including some plated in 22 carat gold, '*the* power statement for the office'). The choice was also between natural spring water, purified drinking water and distilled water.

This is the essential difference between the American and European approach to bottled waters. The European thinks of mineral and spring waters being bottled at source, and may well drink them for their therapeutic benefit. The American is usually more concerned with the purity and sterility of the water than with its pedigree from a historic source. Much bottled water in the US is actually processed tap water or distilled water. The US Food and Drug Administration (FDA) has no official definition of 'mineral water', although in California a water with over 500 mg per litre of dissolved mineral salts must be labelled mineral water and any below may not use that accolade. However, new FDA regulations are expected by the end of 1994. One crucial bone of contention they should resolve is the accepted definition of a 'spring' water. Can the word 'spring' be applied only to water that flows freely from the ground or gushes from a mountainside, or may it be used equally to describe water from an underground aquifer tapped by a borehole? California, which does have the legislation, actually prefers springs set up through boreholes on the grounds that it is a more satisfactory and hygienic way of getting the water out. The word in the industry is that the federal authorities will agree to this broader definition, too. A narrow definition of free flowing from the ground would hit groups like Perrier, half of whose American owned springs have boreholes.

Even if a water is labelled as 'spring water', that is no guarantee in some states that it truly comes from a spring. Indeed, the promotion for a water may herald it as 'natural spring water' while being remarkably vague about the source of that spring. Moreover, a bottler may switch from one spring to another, but still use the familiar name of the original, so that it becomes a brand name. This was a tactic used in launching a water called Ice Mountain (subsequently taken over by Perrier) with the intention of creating a national image for a bottled water from a variety of springs. Perrier has maintained the concept, and is currently pushing Ice Mountain in the Mid-West.

Water is also frequently taken from the spring to the bottling plant by tanker truck – something that is anathema to Europeans. And it is almost always ozonated or subjected to ultra violet light before bottling to render it sterile and bacteria-free. That is precisely what Americans are looking for – a pleasant, pure drinking water free from all traces of pollution that does not have the tang of chlorine afflicting so many municipal supplies, and is free from sodium and other minerals that worry the diet-conscious. In short, the waters are unpretentious; bottlers may not claim the water does you any good, they do guarantee it will do you no harm. The only concession has been the fruit flavours added to 'New Age' bottled waters, which have become very popular.

The addiction to bottled waters, however, remains something of a regional affair, influenced both by climate and social customs. California, that sunny, health-conscious state (also beset by earthquake fears), easily leads the way. Beverage Marketing Corporation's profile of 'Who's Uncapping The Most?' for 1993 showed California

"This isn't tap water, is it?"

and the other west coast states were on 19.4 gallons (73.4 litres), which puts them equal to many European countries, followed by Texas and the south-west on 11.5 gallons (43.5 litres), New York and the north east on 8.6 gallons (32.5 litres) the south (including Florida) at 4.5 gallons (17 litres), with the states of mid-west trailing at 3 gallons (11.3 litres).

The pursuit of that thirst, especially in California, Texas, and from Pennsylvania north-east through New York and the New England states, has shaped an immense re-grouping of the bottled water business in the last decade. In the mid-1980s there were still many small springs and family bottling plants. Today, over half of all sales are in the hands of the top 10 bottled water companies, according to Beverage Marketing Corporation's 1993 statistics. And virtually one quarter are with one group – Perrier. The Perrier Group of America (part of Nestlé Sources International) had 23.2 per cent of all sales in 1993, with a wholesale value of $687 million. 'An iron hand grip on the bottled water business,' observed the magazine *Beverage World*. Only a modest part of this comes from the flagship water itself ($59 million). For the rest, Perrier has gradually taken under its wing many of the major American sources and bottlers (and that diversity helped the group when millions of bottles had to be recalled). The Perrier group began in 1980 with Calistoga, a natural mineral water in the Napa Valley of California and Poland Spring in Maine. Then came a string of acquisitions in the thirstiest states: Arrowhead, long the leader in California; Ozarka, Oasis and Utopia in Texas; Great Bear, an old established water company with springs in Pennsylvania and New York; Zephyrhills in Florida and, most recently, Deer Park, long one of the most famous names in American waters, originally from Maryland, but now also bottled at a spring in Florida.

Perrier's power game, incidentally, has not been followed by its major French competitor, Danone. Their European waters, such as Évian, are sold very successfully in the United States through an offshoot, Great Brands of Europe, which is actually fourth in the American league with sales of $140 million in 1993. But apart from a brief fling with the waters of Saratoga Springs in New York state, Danone has not sought to round up a lot of local American sources. There is a cautionary tale, too, in that Vittel from France (now part of Nestlé) also tried to get a foothold in California a decade ago by re-opening Bartlett Mineral Springs in the Mendochino National Forest north of San Francisco (see the poster onpage 151), once a famous spa, but without success. The lesson is clearly that, to succeed in the US, you either bring in the 'premium' European waters, for which there is a good niche market, or adapt totally (as Perrier has done) to the American taste and mood, offering cheap, well packaged pure local waters by the gallon for the home or office.

This lesson is borne out by the other name players in the American league. The second player is the McKesson Corporation, with around 8 per cent of the market, whose top brand, Sparkletts, rivals Arrowhead for delivery in southern California. Anjou International, in third position, fulfils the same role around Chicago through its subsidiary, Hinckley & Schmidt.

Up against such local competition, Great Brands of Europe, leading with Évian, is doing well to be in fourth place. Overall French waters account for 55 per cent of US imports, although down from 86 per cent some years ago. The new challenge has come from Canada, but that is more in bulk deliveries, as Canadian bottlers, like Naya, exploit their abundant fresh water springs. San Pellegrino seems to be on the table of almost every Italian restaurant in America, easily making Italy the third biggest foreign supplier. All told, however, imports of foreign waters hardly amount to 4 per cent of US sales.

Back on the home front, Mountain Valley Spring Water, famous and nationally distributed long before President Clinton gave it a boost, also retains a special position as a good spring water bottled at source. They are in the top 10 waters, with wholesale sales of $47 million in 1993. Mountain Valley is now part of Sammon Enterprises, which also owns Carolina Mountain Water and Diamond Spring Water. Overall Sammon is sixth in the American league.

Historic collection of American water bottles owned by Austen Hess of Ephrata Diamond Spring, Pennsylvania.

The little leaguers should not be forgotten either. While the last decade has seen a scramble to corner the best-selling waters, America remains a regional market, with plenty of local brands.

Newcomers, however, are few. 'It's a matured market,' said Lisa Prats of the International Bottled Water Association. 'There are not many new brands.' What has really happened is largely a re-shuffling of the old under conglomerates like Perrier, which are taking more waters from abundant springs. When we first went to Poland Spring in Maine in 1984, for instance, only about 9 million gallons (34 million litres) were bottled; in the mid-1990s Poland Spring bottles over 100 million gallons (378 million litres).

Although growth in consumption did slow down in the recession of the early 1990s, water pundits confidently predict that it will double again by the millennium. By the year 2000 Americans could, appropriately, be drinking 20 gallons (75.7 litres) each a year. That is almost up to the present level in France. To think that just 20 years ago, before the Perrier invasion, the United States was considered 'an under-developed country' where bottled water was concerned.

National Association:
International Bottled Water Association,
113 N. Henry Street,
Alexandria,
Virginia 22314, USA.

Mountain Valley: still and moderately carbonated, low mineralization.

MOUNTAIN VALLEY

The still water of Mountain Valley from Hot Springs, Arkansas, is the best-known natural spring water in the United States, and is one of the very few to be marketed throughout the nation. The water, salt-free and mildly alkaline, is chiefly sold in distinctive green glass half gallon (1.5-litre) and 5-gallon (18-litre) bottles delivered direct to homes (a family of four downs up to 20 US gallons a month) and offices. It is also now available in non-returnable plastic bottles, while Mountain Valley Sparkling, a carbonated version of the still spring water was launched in 1990.

Mountain Valley is one of the many springs in and around Hot Springs National Park, whose waters were long renowned in Indian legends for their healing powers. Hot Springs itself blossomed as a spa in the late 19th century, once the railroad arrived. On the main street, nicknamed Bathhouse Row, such visitors as William Jennings Bryan and Theodore Roosevelt took the waters or bathed in Big Iron, Horseshoe or Magnesia springs. Today, most hotels and motels in Hot Springs still offer mineral baths. As for Mountain Valley itself, the Register of the United States noted in the 1890s: 'The purity of the water – having no trace of organic matter – makes it acceptable to the most delicate stomach when nothing else will be retained.' Today most hotels and motels still offer mineral baths, whirlpools and massages to the 1.5 million visitors a year.

Outsize bottle adorns the spring house, where visitors taste and locals collect water.

Mountain Valley itself is 12 miles (19 kilometres) out of town between Glazypeau and Cedar Mountains. The spring is in the midst of 500 acres of forest owned by Mountain Valley Spring Company. Rain falling on the Glazypeau and Cedar Mountains nearby filters through layers of shale, sandstone and limestone into an aquifer 1,600 feet (500 metres) down, from which it then rises through beds of Ordovician marble. Mountain Valley claims that geological tests have shown this process takes 3,500 years. Visitors to Mountain Valley can watch this ancient vintage bubbling out, beneath a small glass dome, at a steady 50 US gallons (189.25 litres) a minute at a constant 65°F. (18°C.). When it first emerges, the water has a distinctive iron flavour, but after being filtered and stored in holding tanks for 72 hours, much of the iron is removed. Mountain Valley is then ozonated.

The water has been bottled regularly since 1871, but sales really took off after a chance meeting on a train between publisher William Randolph Hearst and gambler Richard Canfield. Both produced bottles of Mountain Valley from their luggage to drink at dinner, agreed they were addicted, and promptly set up a joint venture in New York to market it. For many years Mountain Valley operated as a consortium owned by its distributors under the chairmanship of John Scott, a tireless advocate of the virtues of natural spring waters. However, in 1987, it was purchased by Sammons Enterprises, Inc., a privately owned company in Dallas, Texas, which also owns two other waters, Carolina Mountain Water and Diamond Spring Water.

Mountain Valley's admirers have included everyone in show business from Gloria Swanson and Frank Sinatra to Elvis Presley and Mick Jagger. President Reagan took it along on all his trips abroad, and President Bill Clinton, whose hometown is Hot Springs, Arkansas, grew up drinking Mountain Valley spring water, and still drinks it.

Production: 9.1 million US gallons (34.4 million litres) annually.

Bottling company:
Mountain Valley Spring Company,
150 Central Avenue,
PO Box 1610,
Hot Springs National Park,
AR 71902,
USA.

Analysis:	*mg/l*
Calcium	*68.00*
Magnesium	*8.00*
Sodium	*2.80*
Potassium	*1.00*
Fluoride	*0.25*
Iron	*0.01*
Zinc	*0.01*

Mountain Valley's source at the heart of a 500 acre Arkansas forest.

Poland Spring: still and carbonated, very low mineralization.

POLAND SPRING

The clear, cool waters of Poland Spring, tucked away amid the woods and lakes of Maine, have been known for 150 years. But their distribution on a grand scale, in still and carbonated form, took place only from the early 1980s, after the spring came under the umbrella of Perrier's American company, then called Great Waters of France. Today Poland Springs is marketed throughout New England, where it is the leading brand, promoted under the slogan 'What it means to be from Maine'.

The legends surrounding the spring's original discovery centre on the Ricker family who owned an inn at South Poland, Maine, from 1797. Various family members found the local spring helped their ailments, but it was not until 1844 that Hirman Ricker began to tout its virtues for easing his chronic dyspepsia. Soon the whole neighbourhood vowed Poland worked wonders for them too. Ricker took to advertising that the water 'Cures Dyspepsia! Cures Liver Complaints of Long Standing! Cures Gravel! Drives out all humors and Purifies the blood'. He built a granite and marble spring house with a mosaic floor inlaid with the words Sapienta Donum Dei (Wisdom is the gift of God). Alongside he installed a bottling plant, surmounted by a tall tower, which was to be used for over a century. And Poland Spring water was soon being widely distributed. A little way off in the woods Ricker then put up a five-storey 100-room resort hotel with gilded turrets and domes. Before long he added a golf course. American and European socialites came to Poland in droves in the late 19th century, and for a while it rivalled Saratoga Springs as a fashionable spa. But the age of spas declined, and so did Poland's fortunes. The Ricker family, however, kept on bottling until 1946, when they sold the spring. Thereafter for nearly 30 years less than 100,000 cases were bottled annually.

The growing bottled water boom, however, suddenly provided a fresh lease of life. Perrier's Great Waters of France, eager to get a foothold with a local American water, purchased the spring, and opened a modern bottling plant in 1980 downhill from the original spring house. The original Poland Spring had a modest flow of only 8 US gallons (30 litres) a minute, emerging from the granite hillside. But Perrier soon located other nearby springs in the surrounding pine forest, with waters of identical composition. They come from an aquifer layer into which they are filtered through very fine sand and gravel left by glaciers 10,000 years ago. Although the water is exceptionally pure and has a very low mineral content of 46 mg/l, it is filtered for sediment and passed before ultra-violet light to eliminate any potential bacteria.

Poland Spring is packaged in 6-gallon, 2.5-gallon and 1-gallon plastic containers, together with three sizes of PET bottles for still and glass bottles for sparkling. Cherry, lemon, lime and orange flavours are also marketed.

Production: Over 100 million US gallons (378 million litres)

Bottling company:
Poland Spring Bottling Company,
PO Box 499,
Poland Spring,
Maine USA.
Owned: The Perrier Group of America, part of Nestlé Sources International

Analysis:	*mg/l*
Sodium	*3.00*
Magnesium	*1.60*
Nitrates	*0.50*
Sulphates	*2.05*
Bicarbonates	*24.00*
Iron	*0.05*

EPHRATA DIAMOND

Driving west from Philadelphia across lush, rolling farmland you arrive after an hour or so in the small town of Ephrata in the heart of Pennsylvania Dutch Country. Ephrata was first settled by immigrants who were attracted by its fine spring water that is naturally filtered through a sandstone ridge, unique among the limestone formations of the area. The reputation of its waters soon made Ephrata a well-known spa. Thousands travelled great distances to put up in large clapboard hotels to 'take the cure'.

Ephrata Diamond Spring Water is a family firm founded almost a century ago in 1895. Initially its water was sold in 10-gallon cans in the nearby town of Lancaster. In 1944 the spring was bought by Edwin Hess; today it is run by his son, E Austin Hess, as chairman of the board and grandson Richard Hess (who was president of the International Bottled Water Association 1993/94). Ephrata Diamond is a lightly mineralized water. Sixty per cent of its sales are through direct deliveries to home and offices, but it is also available at grocery stores in 10 states.

GREAT BEAR

This is a natural spring water, widely distributed in Washington, New York and the north east. Its name derives from a spring in up-state New York used by Iroquois Indians, who called it **Great Bear** after a legendary tussle they fought with a gigantic polar bear. Great Bear has been bottled since 1888. Today the waters of the Great Bear Spring at Washingtonville in New York state and of another spring in Pennsylvania are bottled under this label.

SARATOGA SPRINGS

Saratoga Springs in New York state has an historic reputation for mineral waters and horse racing. Colourful characters from real life, like Diamond Jim Brady in the 1890s, and fiction, such as James Bond in *Diamonds are Forever*, paraded there. The initial attraction, before the horses, was an abundance of over one hundred mineral springs, both saline and alkaline, in and around the town. Among the most famous is High Rock Spring, bubbling from a pyramid of calcerous deposits built up over centuries. The waters come from a layer of dolomite limestone, overlain by shale. Several of the more highly mineralized ones have long been used for baths. Traditionally, the waters that were bottled were compared with those of Vichy in France, labelled as Saratoga Vichy Water.

Today, the Saratoga Spring Water Co, presents a more modern image with lightly mineralized sparkling and still waters and New Age beverages. They are distributed down the east coast into Tennessee and were being launched in California in 1994. The town has also brushed itself up. "The old spa buildings have been remodelled, "Saratoga Springs' Pat Rogers told us, "we've got outdoor cafés - the whole place has a European flavour." Just like Vichy, in fact.

Georgia Mountain Water: still, low mineralization.

GEORGIA MOUNTAIN WATER

Visitors to the Carroll family's 200-acre farm in the Blue Ridge Mountains of northern Georgia were long refreshed with glasses of cool, clear water from local springs. The mountains around the farm form part of the Blue Ridge Divide from which water runs north towards the Tennessee river and south towards the Coosa river so that no water from any other property drains across the recharge area for their springs.

But the water was used only for local campers until Jim Carroll, already a successful businessman and now president of Georgia Mountain Water, read a magazine article on the benefits of bottled waters. The prospect of marketing his farm's spring waters seemed tempting. In search of improved flow, he drilled down 165 feet (50 metres) through granite beside the spring and discovered an abundant underground source of pure water with exceptionally low mineralization - a mere 26 mg/l. So Georgia Mountain Water was launched in 1982, very much as a family affair. Jim Carroll runs it; his brother Steve looks after distribution; and brother Tom manages the computerized bottling factory.

This plant, completed in 1989, enabled Georgia Mountain Water to enter the big league of natural spring waters. Sales soared from 0.5 million US gallons (1.9 million litres) in 1988 to 5 million (19 million litres) by 1994. The water is sold in recyclable plastic bottles from 12 ounces (354ml) to 50.7 ounces (1.5 litres) in Atlanta and throughout much of the south eastern United States; it is also available in 5-gallon (19.375-litre) plastic bottles for direct delivery to homes and offices.

The marketing has been simple; this is a clear, pure mountain water with 'no chemical additives, sodium, preservatives, sweeteners, flavourings, artificial gases, and best of all – no calories'. Just what the majority of American bottled water drinkers are seeking.

In researching this new edition and sorting out the conglomerates that have grown up in the US water business over a decade, it was pleasing to find that the name first recommended to us to include as a 'newcomer' was a family business thriving on a simple mountain spring water.

Production: 5 million US gallons (19 million litres).

Bottling company:
Georgia Mountain Water Inc.,
PO Box 1243,
Marietta,
Georgia 3006, USA.

Analysis:	*mg/l*
Calcium	*1.700*
Fluoride	*0.014*
Iron	*0.010*
Magnesium	*0.470*
Potassium	*0.600*
Zinc	*0.010*

FLORIDA

Florida's warm climate and its people, many retired or holiday, have always been in favour of bottled waters. One of the first enthusiasts we encountered a decade ago was Bob Levin, the Miami distributor for Mountain Valley spring water, who had organised *Les Amis des Eaux* (Friends of Water), which held regular tastings. The best-selling water is **Zephyrhills**, from the small town of that name in central Florida between Orlando and Tampa, which calls itself 'The City of Pure Water'. The waters of a deep spring at Zephyrhills have been bottled since 1969; since 1987 it has been part of the Perrier Group of America. While being top water in Florida, it also ranks eighth in the US, with sales of just under $50 million in 1993. Also in Florida the waters of 'The Spring of Life' at Monte Verde are sold under the **Deer Park** label (now also in Perrier's empire).

TEXAS

Texans, not surprisingly, have a great thirst, second only to those living in health-conscious California. They get through around 12 US gallons (45 litres) per head annually, almost equal to continental European levels. Three Texan waters, **Ozarka, Oasis** and **Utopia** (all owned by the Perrier group) cater for their needs. **Ozarka**, which is the best-seller, comes from the Moffit spring in Nacogdoches County between Houston and Dallas which was once the major source of water for Indians in the area. It has been bottled since 1905 and is available in both still and sparkling. Perrier acquired it in 1987. **Oasis**, in the water business since 1946, offers the usual mix of spring, distilled and drinking water. And **Utopia**, rather a newcomer, started bottling the waters of a spring, once known by the Indians as 'Laughing Water' at Utopia in the hill country of Texas in 1983. Its spring water, together with drinking and purified water, sells mainly in San Antonio, Houston and south Texas.

Calistoga: carbonated, medium mineralization.

CALISTOGA

Calistoga mineral water is bottled at the source in the small town of Calistoga, 75 miles (120 kilometres) north of San Francisco, in the famous wine-growing region of California's Napa Valley. With 3,900 inhabitants, Calistoga boasts eight establishments offering hot sulphur baths and thermal treatments to visitors.

The hot springs of the Napa Valley region were known to the Indians and the first Spanish settlers of California. Legend has it that the first enthusiast from the east coast wanted to rival Saratoga Springs, located in New York State, and announced: 'I will make this the Saratoga of California'. Or at least, that is what he thought he said. What actually came out was: 'I will make this the Calistoga of Sarifornia'. Anyway, the name stuck.

In August 1920, an Italian immigrant escaping from the San Francisco earthquakes bored the 'Guiseppe Musante Hot Sulphur Water Geyser' and production started at the bottling plant in 1924.

A mineral water, as defined by Californian law, contains more than 500 parts per million total dissolved solids; and Calistoga water contains 547 tds. Calistoga water emerges from a hot geyser at 212°F (100°C), and is cooled to 51.8°F (11°C) for bottling. Calistoga bottling company has a well 230 feet (70 metres) deep, with a pump pulling from 110 feet (33 metres).

Care has to be taken not to pump the water too fast or it will vaporize; water is in fact pumped at 70 US gallons (264 litres) a minute. After cooling, the water is filtered to eliminate sediment, and then filtered through green sand.

Calistoga is the biggest domestic mineral (as opposed to spring) water in the United States. Seventy per cent of sales are in northern California which, the general manager reminded us, is 'the most health-conscious state'. But it also sells in Washington, Oregon, Arizona, and as far away as Hawaii. A still spring water and flavoured varieties are also marketed.

Production: 15.9 million gallons (60 million litres).

Owned: The Perrier Group of America, part of Nestlé Sources International.

Bottling company:
Calistoga Mineral Water Co. Inc.,
1477 Lincoln Avenue,
Calistoga,
California 94515, USA.

Analysis:	*mg/l*
Sodium	*150.0*
Potassium	*16.0*
Calcium	*7.0*
Magnesium	*1.0*
Bicarbonates	*130.0*
Nitrates	*0.2*
Fluoride	*9.5*
Iron	*0.1*

CRYSTAL GEYSER

Leo Soong and Peter Gordon were investment brokers at Wells Fargo Bank when they decided to start a business in line with their own physical fitness interest and shifting market trends – mineral water. They searched California's Napa Valley, tasting almost 40 different springs, until they found the one that they judged tasted best when carbonated, on the southern edge of Calistoga town, an area famous for its hot geysers.

The spring which Soong and Gordon bought in 1977 and named Crystal Geyser, erecting a bottling plant above it, emerges from an aquifer 240 feet (73 metres) below the earth at a temperature of 140°F (60°C). The factory cools the water, filters it free of sediment, and then ozonates and carbonates it before placing it in 10-ounce and 1-litre recyclable glass bottles.

From the start, great emphasis has been put on promoting the water to sports fans by sponsoring athletics and bicycle races. Crystal Geyser is a sponsor of the San Francisco Bay to Breakers Race, the Los Angeles Marathon and the Squaw Valley Bike Race. It also supports the San Francisco Symphony Orchestra, the San Francisco Opera, the American Parks Network and public television broadcasting. 'It doesn't hurt us that we do well with the people with better taste buds,' says Leo Soong. 'Palate in the end is the deciding factor.'

To reach more palates, the company also introduced a range of drinks based on its sparkling mineral water and flavoured with a variety of natural fruit essences, such as cola berry and wild cherry. It markets other drinks based on combinations of exotic fruit juices and the original mineral water, some with added vitamins, but all without artificial colours, flavours or sweeteners.

Catching the vogue for still water, Crystal Geyser launched its alpine spring water in 1990. The source is at 4,000 feet (1,223 metres) on the slopes of Olancha Peak, near Mount Whitney in California's high Sierra. The snow from these peaks takes up to a century to filter through the rock from the Crystal Geyser source. The success of this water, bottled at source, won it the 'Bottled Water Big Splash' award in 1992 for 143 per cent growth in the value of sales to $28 million, helping to push the Crystal Geyser group to eighth place in America's top ten bottled water companies.

Crystal Geyser Sparkling Mineral Water: carbonated, medium mineralization.

Crystal Geyser Alpine Spring Water: still, low mineralization.

Bottling company:
Crystal Geyser Water Company,
PO Box 304, Calistoga,
Napa Valley,
California 94515-0304.
Part of the Otsuka America group.

Crystal Geyser Sparkling Mineral Water:	*mg/l*
Sodium	160.0
Calcium	8.1
Magnesium	2.8
Sulphates	2.6
Chlorides	260.0
Bicarbonates	170.0
Fluoride	7.0
Iron	0.1

Crystal Geyser Alpine Spring Water:	*mg/l*
Bicarbonate	97.5
Chloride	24.4
Fluoride	6.5
Magnesium	6.0
Nitrate	0.6
Potassium	2.0
Sodium	13.0
Sulphate	36.7

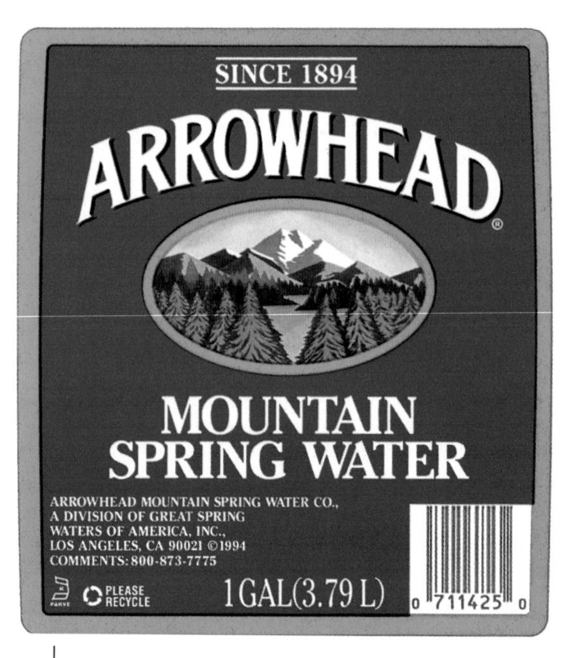

CALIFORNIA

Californians are the champion drinkers of bottled waters in America at around an annual 20 gallons (75.6 litres) per head. The state also pioneered the concept of in-store water bars with names such as H_2O selling a range of mineral waters and hundreds of copies of the original edition of this guide. California also had its own tradition of famous spas in the 19th century, including Bartlett Mineral Springs, whose waters were bottled by Vittel a few years ago.

Among the state's widely sold waters is **Arrowhead Mountain Spring Water**, which has been bottled for 100 years, is the best selling water not just in California, but in the United States, with wholesale sales of $220 million in 1993, being 7.4 per cent of all American bottled water sales.

The original Arrowhead spring is high in the San Bernardino mountains east of Los Angeles. The legend has it that Indians found the spring after noticing a natural rock formation in the shape of an arrowhead pointing towards it. The water is now taken from a number of springs in southern California and over half its sales are direct delivery, often in 5-gallon containers to homes and offices.

Black Mountain prides itself on being the largest independent spring water business in the United States, ranking 10th in the league. The source was discovered by George Washington Faulstich riding near San Carlos 25 miles south of San Francisco in 1937; he christened it Black Mountain because of the deep, dark shadows cast on the nearby hills by the setting sun. Black Mountain remains in the family and is distributed widely in San Francisco and the Bay area, where it is the 'official' water for football and baseball teams.

Mendocino: naturally lightly carbonated, high mineralization.

MENDOCINO

Mendocino is one of the most highly mineralized waters being bottled in the United States and one of the few that is naturally carbonated. While most American waters underline what they do not contain, Mendocino prides itself on a rich blend of sodium, magnesium and bicarbonates. The total dissolved solids count is more than 1,500 mg/l.

The water comes from a naturally carbonated artesian spring, deep in the redwood forests north of San Francisco, at a constant 53°F (11.6°C) with a good flow of 60 gallons (227 litres) per minute. The natural carbon gas gives it a modest effervescence, like Bru in Belgium or Ferrarelle in Italy. The spring was discovered in the late 1870s and the water was sold as 'G M Henderson's Bonanza Mineral Water' for 30 years. However, after the area became a lumber camp around 1900 all bottling ceased.

A San Diego couple, Ty and Gayle Hill, revived bottling in 1982 after buying the ranch on which the spring is located. Analysis revealed a pure, highly mineralized water, so they put in an underground stainless steel pipeline to a small new bottling plant among the redwoods. The water passes through a sand filter and a small charcoal bed, and is ozonated.

Mendocino is now owned by Mendocino Beverages International of Santa Rosa, which distributes it as an 'Estate Bottled' mineral water and in a range of fruit flavours throughout the West Coast, mainly to natural food stores, but also to restaurants, where it is in increasing demand as a unique water.
Production: 1.5 million gallons (5.7 million litres).

Bottling company:
Mendocino Beverages International Inc.,
200 Talmadge Drive,
Santa Rosa,
California 95407,
USA.

Analysis:	*mg/l*
Sodium	*240*
Calcium	*310*
Magnesium	*130*
Bicarbonates	*940*
Chlorides	*69*

CANADA

INTRODUCTION

Canada has a natural abundance of good water from springs in its vast wilderness of lakes, forests and mountains. These sources place it in a unique position; they offer not only supply for the local market but ample waters for the huge American market over the border. Although Canadian consumption has risen three-fold over the last decade, some Canadian bottlers have deliberately chosen almost to ignore home ground, and have focused on slaking American thirst.

Thus **Naya** from an aquifer in the Laurentian mountains near Mirabel, first tapped in 1986 by four Lebanese investors, sells over 80 per cent of its water directly into the United States (and is pressing on Mexico and other Latin American destinations and Japan). The cool northern water appeals down south. 'We have a good image in export markets,' says Louise Bouchard, president of the Association des Embouteilleurs d'Eau du Quebec.

On the home front, the Canadians themselves are drinking around 400 million litres (105 million US gallons) a year, but that is a modest 15 litres (3.9 US gallons) per capita, or less than half of American consumption. Still water has over 90 per cent of the market. The climate, of course, does not help. The harsh Canadian winter does not do wonders for bottled water sales. Moreover, the story of water in Canada is somewhat akin to its politics – Quebec against the rest. The French speaking province not only has some of the best sources, but shares the French tradition of appreciating mineral waters (and 65 per cent of Perrier's Canadian sales are in Quebec). Quebec still accounts for close to 80 per cent of waters bottled in Canada. It is the only province with strict government regulations insisting on bottling at source. The Quebec code, modelled on French regulations, has long ruled that *l'eau minérale* (mineral water) must contain at least 500 mg/l of dissolved minerals and, if sparkling, must state whether naturally or artificially carbonated.

Quebec's predominance is clear the moment you cast an eye down the top 10 companies bottling water in Canada; six are in the province, and that is without including a wide range of other waters, like **Saint Justin**, a carbonated natural mineral water bottled since 1895. Two sparkling waters from the springs of St. Joseph des Cédres and Source St. Lazare are bottled under the **Montclair** label, which is part of Nestlé Sources International, and sell mainly in Quebec. A still water from the Piedmont spring is also sold by Montclair, but mainly in the English-speaking provinces. Quebec City itself has long been the terrain of **Boischatel** and **Cristalline**. Overall in the province, **Labrador** is the best selling still water, vying against **Laurentian Spring Valley**. Labrador used to be part of the Nestlé group, but since 1992 has been owned by Aquaterra, which also owns **Crystal Springs** in Ontario; the Aquaterra group is now number one in the overall Canadian league.

Despite the traditional Quebec taste for mineral waters, the trend there, as in the rest of Canada and the United States, has been towards waters with lower dissolved solids. Laurentian, for instance, switched from its traditional spring which had 400 mg/l dissolved minerals to two others with a modest 160 mg/l. The big business, too, is in delivery of bottles of 18 or 22 litres to the home and office, which accounts for 56 per cent of the Canadian market.

Beyond Quebec, however, the bottled water business is growing up. As Quebec Association president Louise Bouchard told us, 'There has been a tremendous increase in the range of spring waters in other provinces. Spring waters were non-existent in provinces like British Columbia 10 years ago.' The Canadian Bottled Water Federation was formed in Toronto in 1993 to preside over the industry nationally. And pending new federal government regulations on bottled waters, has set up its own Bottled Water Model Code, which provides for independent inspections of members' sources and bottling plants. It also provides a definition of 'spring water' as 'potable water derived from an approved underground source that contains less than 500 mg/l total dissolved solids, but not obtained from a public community water supply'; while 'mineral water' must fulfil the same criteria, but with not less than 500 mg/l dissolved solids.

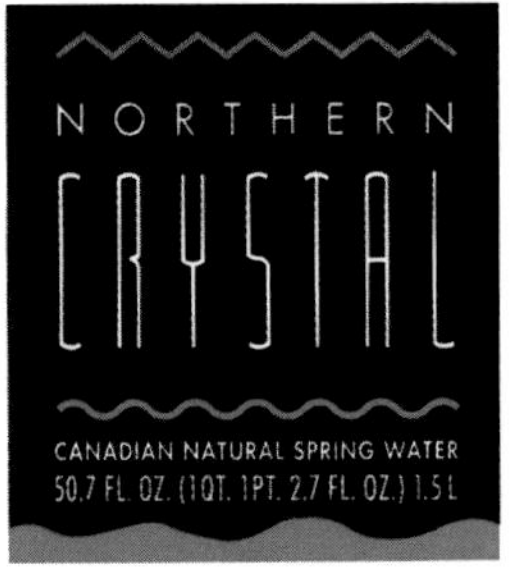

The province of Ontario is also the second market for bottled waters. Crystal Springs, part of Aquaterra, has been the best seller there for years. The water comes from the Cataract source in the Caledon Hills, near Niagara Falls. Cataract is surrounded by 67 acres of virgin land backing onto a provincial park, and the water is in an aquifer overlaid with clay and shale. The water was first bottled in 1883 and was originally used by Canada Dry for its ginger ale and soft drinks. Since 1963, however, it had been bottled as a natural spring water. Crystal Springs has modest mineralization, with a touch of sodium (17.3 mg/l), magnesium (17.3) and calcium (87.6), with total dissolved minerals at 368 mg/l. Although sold primarily in Ontario, marketing across Canada began in 1994. An export brand, **Northern Crystal**, is sold in the US and the Caribbean.

While Quebec and Ontario contribute most drinkers, the maritime province of Nova Scotia has some good springs, such as **Sparkling Springs** from Dartmouth, which ranks seventh in the Canadian sales league and the intriguing newcomer by the name of **Canada Geese** (see next page).

Out west the choice is more limited, though if you are passing through Calgary, Alberta, you may taste **Vivant Natural Spring Water**, or in Edmonton try **Sierra Spring**. While in British Columbia, **Canadian Spring Waters** makes most of the running – they are sixth in the Canadian league.

Although Canadian waters are largely regional, both Évian and Perrier maintain national distribution and command 'premium' water sales.

National Association:
Canadian Bottled Water Federation,
30 West Beaver Creek Road, Unit 7,
Richmond Hill,
Ontario L4B 3K1,
Canada.

Quebec Association:
Association des Embouteilleurs d'Eau du Quebec,
407 boul. St. Laurent,
Bureau 500,
Montreal ,
Quebec H2Y 2Y5,
Canada.

Canada Geese: still and lightly carbonated, high mineralization.

CANADA GEESE

Canada Geese is a new Canadian mineral water from a very old source – a reservoir beneath the wooded Annapolis Valley between the North and South Mountain ranges of the Nova Scotian peninsula. At Spa Springs (formerly Wilmot Spa), near Annapolis Royal, the first capital of the Province, the water bubbles spontaneously to the surface, cold and clear, from a well 55 metres (180 feet) deep. Scientific tests have shown that the water is at least 3,500 years old – when the Pharoahs ruled Egypt Canada Geese first fell as rain and began to percolate through the land surrounding Spa Springs. Granitic rock formations through which it flows have given the water its high mineral content, and protected it from pollution and contamination.

Canada Geese is a calcium-sulphate water, which has been found beneficial for skin and digestive complaints, rheumatism and kidney problems. Its therapeutic properties were recognised by the Micmac Indians who inhabited the area in the 16th century, and by subsequent settlers. In 1817, the owner of the springs was persuaded by an English visitor to build a small bottling plant, and to erect two wooden buildings for hot and cold baths – and Wilmot Spa was founded. By the 1880s the beautiful spa town had become very fashionable and was attracting visitors from all over Canada and the United States.

In 1891 a new bottling plant was built to package and export the Spadeau mineral water (as it was then known). A year later a second hotel was opened to replace an earlier one destroyed by fire, and in 1894, the spa received the distinction of being mentioned by Karl Baedeker in his first German guidebook for Canada. After the turn of the century, the fortunes of Wilmot Spa declined.

Ninety years on, there has been a renaissance at Spa Springs. The land was bought in the late 1980s by the present owners, Spa Springs Parks Ltd, from the Elliott family, which had acquired it in the 1930s and carefully preserved the mineral water springs. Hydrogeological explorations were carried out, and a brand new, fully automated bottling plant was completed in late 1987, all with the constant advice of the respected Institut Fresenius in Germany.

The brand name Canada Geese was chosen because these birds symbolize elegance, strength and health: 'All attributes', comments Dr Vladi, a major shareholder in the production company, 'which can so easily be associated with our mineral water.' Two versions of Canada Geese are available – still and gently carbonated. The latter has 2 grams of CO_2 per litre added.

Initially, Canada Geese was distributed mainly in eastern Canada and Quebec, with small quantities also sold in Boston and New York. Once Canada Geese has been approved as a natural mineral water in Europe and is selling there, the company will begin the official US launch.

Production: approximately 3.5 million litres (0.9 million US gallons).

Bottling company:
Canada Geese,
Mineralwater
Company,
Spa Springs,
Middleton,
Nova Scotia,
Canada.

Analysis:	*mg/l*
Sodium	*36.00*
Potassium	*1.95*
Magnesium	*9.60*
Calcium	*282.00*
Fluoride	*0.22*
Chloride	*8.30*
Nitrate	*2.00*
Sulphate	*688.00*
Hydrogen carbonate	*119.00*

Naya: still, low mineralization.

NAYA

As a success story, Naya natural spring water from the Laurentian mountains of Quebec, takes some beating. The spring, located after five years of exploration for a suitable water, has been bottled only since 1986, but is already not just a leading seller in the 'premium' still water market in Quebec (edging out the imported Évian), but is virtually the only Canadian water with national distribution. That is hardly the beginning; Canada actually accounts for under 20 per cent of sales; almost 80 per cent of the bottles go to the United States, where Naya is now the fourth best-selling imported water (in still water it is second only to Évian). It is also exported to Mexico and the Caribbean, to Japan and the Middle East. Naya is a Canadian water that thinks international. 'From day one our dream was export, export, export,' Naya's director, Naji Barsoum, told us.

Naya is owned by Nora Beverages Inc.in Mirabel, Quebec, a company set up by four Lebanese investors led by Ahmed Hbouss. The spring is named after the mythological Naiads, the Greek deities who were the legendary protectors of water flowing in springs, rivers and fountains. The Naiads brought fertility and growth. The water comes from an aquifer at a depth of 250 feet (76 metres); it is overlaid with layers of sand and gravel, stratified clay, limestone, shale and compacted sand. It has light mineralization with total dissolved solids of 215 mg/l and is bottled at source in sizes of 1.5 litres or less (unlike most Canadian waters which are packaged in larger containers for home or office delivery). Thus Naya is much closer to the tradition of many European still waters, which it is challenging in North American and other export markets.

Production: over 30 million litres (7.2 million US gallons).

Exports: Over 80 per cent of production to United States, Mexico, Japan.

Bottling company:
Nora Beverages Inc.,
2500 Naya,
Mirabel, Quebec J0V 1Z0,
Canada.

Analysis:	*mg/l*
Calcium	38.00
Magnesium	20.00
Sulphates	15.00
Sodium	6.00
Potassium	2.00
Fluorides	0.14

BRAZIL

INTRODUCTION

'Brazil is a country rich in mineral waters,' notes Dr Ruy Bueno de Arruda Camargo in the introduction to his *Águas Minerais Brasileiras*. He then goes on to chronicle the characteristics and potential therapueutic benefits of no less than 293 mineral water springs in 96 thermal stations. Indeed, outside Europe, Brazil presents perhaps the most highly organized and best regulated mineral water industry to be found anywhere. And the sheer range of mineral content and temperature at which the waters emerge is also hard to equal. One source, Bom Jardim, has 5,500 grams mineralization per litre, while Fonte do Bambual at Lagoa de Pirapitinga bubbles out at a scalding 57°C (135°F). Several are naturally carbonated.

Although many of the waters may be taken only under medical supervision at spas which still thrive throughout Brazil, 93 mineral waters are bottled as acceptable '*águas minerais de mesa*' (table waters). All are registered and controlled by the Departamento Nacional de Produção Mineral (DPNM), which leases the water rights on each source to the bottling companies, and charges an annual tax. Production, which was on a plateau during the late 1970s and early 1980s as Brazil's economic difficulties stilled growth, has doubled over the past decade to around 1 billion litres (264 million US gallons) as demand soared.

Still waters are slightly more popular than sparkling, although much of the still market is in 20-litre bottles sold by such bottling companies as Indaia Águas Minerais for office water coolers. But the naturally carbonated ones, Caxambu, Cambuquira, Lambari and São Lourenço, from deep volcanic sources of Minas Gerais province, have carved out a significant niche, especially in Rio de Janiero. Their light effervescence goes down well on the Copacabana and Ipanema beaches.

Mineral waters are found throughout Brazil (even near Manaus in the Amazon jungle), but the greatest concentration of famous spas are within a few miles of each other in the south of Minas Gerais province, where there are 40 sources, most of volcanic origin, and in São Paulo province, which has 100 sources, providing almost 40 per cent of all output. The best-known names are São Lourenço, Caxambu in Minas Gerais, and Lindoya and Campos do Jordao in São Paulo. These towns pride themselves as health resorts with hotels, drinking halls, spacious parks and special facilities for baths and massages. And the potential for marketing their waters has lured such multinationals as Nestlé Sources International (formerly Perrier), and Brazil's Supergasbras and Edson Queiroz, whose main activities are in the gas production and distribution business. Perrier controls São Lourenço together with one important spring, Levissima, at Lindoya and Petropolis, an important still water source near Rio de Janiero. The Brazilian natural resources group Supergasbras has joined the increasingly competitive fray, buying up not only Caxambu, but three other Minas Gerais springs, Araxa, Cambuquira and Lambari.

The take over by large groups, all installing up-to-date bottling lines, of what were previously local mineral water springs clearly signals a change in attitude in Brazil. Although many of the waters have been drunk for their therapeutic benefits since the mid-19th century, the prime motive for many people has simply been that bottled water was safer than tap water, especially in the major cities. A point underlined by Indaia Águas Minerais, who are the market leaders through selling 20-litre bottles to offices and homes. 'During the last 10 years consumption has doubled. The main reasons are the increase in population and the substitution of bottled water for tap water, says Nikita Droin, director general of Nestlé Sources International in Brazil. 'Until now, Brazilians have not been drinking "health" but "purity". Our goal is to make the consumer conscious of the added value of Brazilian mineral waters.'

National Association:
Rio de Janeiro
Refrescos SA,
Estrada do. Itarare,
Rio de Janeiro 1071,
Brazil.

Lion's head fountain in Parque das Áquas at São Lourenço.

Fonte Oriente: naturally carbonated, low mineralization.

Fonte Andrade Figueira (São Lourenço sem gas): still, low mineralization.

SÃO LOURENÇO

The subtle natural carbonation of the waters of Fonte Oriente at São Lourenço makes it one of the most refreshing of Brazilian waters, somewhat akin to Badoit in France or Ferrarelle in Italy.

The mineral waters of São Lourenço in the valley of the Rio Verde in Minas Gerais province midway between Rio de Janiero and São Paulo, were discovered in the early 19th century. But São Lourenço was not developed as a real spa until 1890 when Empresa de Águas São Lourenço SA was formed to establish a resort and market the waters.

There are six warm springs of deep volcanic origin, coverering a wide range of mineralization. Fonte Oriente, emerging at 19.4°C (66.9°F), is the mildest and most widely commercialized. It is billed as a good diuretic and digestive water. Fonte Andrade Figueira, known as São Lourenço sem gas, is also bottled. The remaining four springs which are now capped but not bottled commercially, are Vichy and Alcalina, both rather alkaline; Primavera, which is rich in magnesium, iron, potassium and sodium; and Jaime Sotto Maior, which is sulphurous.

All the waters may be sampled in little white pavilions spaced out among the trees and lawns of the *Parque de Águas*, beside the shores of a clear, cool lake. And they are used in a wide range of treatments at a modern centre for hydrotherapy. Indeed, São Lourenço, like Contrexéville or Vichy in France, is trying to cut a dashing image as a resort for health buffs with golf, boating, roller-skating and swimming. So it comes as no surprise to find that, since 1974, the spa and its waters have been controlled by the Perrier group (now part of Nestlé Sources International) from France.
Production: 13 million litres (3.5 million US gallons) annually.

São Lourenço Fonte Oriente:	*mg/l*
Calcium	*67.9*
Magnesium	*65.0*
Potassium	*39.5*
Sodium	*81.8*

Bottling company:
Empresa de Águas São Lourenço SA,
Praia de Botafogo 228, Suite 713,
Rio de Janeiro RJ,
Brazil.
Owned: Nestlé Sources International.

Caxambu: still and naturally carbonated, low mineralization.

CAXAMBU

The naturally effervescent waters of Caxambu in southern Minas Gerais were known to the Indians long before the Portugese settled in Brazil. They were soon christened *Bolhas a Ferver* (boiling bubbles), and by the early 19th century these medical benefits for kidney and digestive ailments were being praised. By 1875 the state of Minas Gerais authorized the first commercial development of the waters, and Caxambu became a popular spa. There are 12 *fontes* (springs) with alkaline warm waters of volcanic origin emerging at about 24°C (75°F). The flow from most is too limited to be bottled commercially, and they are reserved for visitors to the *Parque de Águas*. But four *fontes* are bottled. Don Pedro, Mayrink I and Mayrink II have light natural carbonation, while Mayrink III is bottled as a still water. The sparkling waters, which have a very clean flavour with a slightly alkaline after-taste, are more popular.
Production: 8 million litres (2.11 million US gallons).
Bottling company:
Superágua, Rua Sao Jose 90 17 andar, Rio de Janeiro, Brazil. Owned: Supergasbras.

Caxambu:	*mg/l*
Calcium sulphate	7.5
Calcium bicarbonate	149.6
Magnesium bicarbonate	89.6
Sodium bicarbonate	67.2
Potassium bicarbonate	54.7
Potassium chloride	6.3
Potassium nitrate	1.3

LINDOYA

This is one of the many mineral springs in the state of Sao Paulo. As the name suggests, it is situated in the vicinity of the towns of Lindoya and Águas de Lindoya, around which a tourist attraction has been built. Owing to its low mineralization, Lindoya has earned the name of *levissima* (lightness).
Bottling company:
Empresa de Águas São Lourenço SA, Sitio Monte Alegre, Bairro do Pelado Águas de Lindoia SP, Brazil.
Owned: Nestlé Sources International

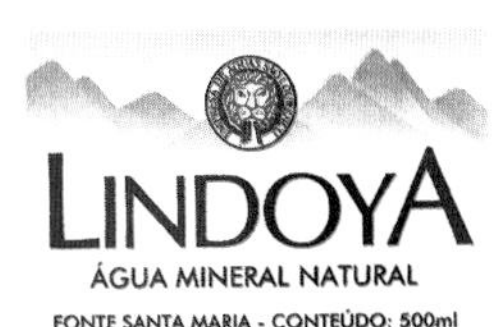

Lindoya: still, low mineralization.

PETROPOLIS

Petropolis is a mineral water spring with close links to the Imperial City, located less than 50 kilometres (30 miles) from Rio de Janeiro. The still water is well known to Rio residents as it is distributed mainly in the state of the former Brazilian capital. Its lightness and very low mineralization means that it is particularly suitable for pregnant women and babies.
Bottling company:
Empresa de Águas São Lourenço SA, Rua Eng. Jose Lima Filho 239, Petropolis RJ, Brazil.
Owned: Nestlé Sources International.

Petropolis: still, very low mineralization.

ARGENTINA

INTRODUCTION

Argentina's vast diversity of geography and climate has produced many famous mineral springs and water sources. These are found throughout the country – in the the towering Andes mountains of the north-west, the grassy pampean plains and the central Patagonian plateau. Some examples of thermal springs are Termas de Reyes in Jujuy, Villavicencio and Los Molles in Mendoza, and Copahue in Neuquen. However, the practice of bottling and selling natural mineral waters on a commercial basis is comparatively recent, and dates back to the earlier part of this century. Villa del Sur, the market leader, which sells throughout the country, was launched in 1971.

Over the past three years, consumption of mineral water has doubled in Argentina to 330 million litres (87 million US gallons) a year. Like the British, Argentinians drink on average around 10 litres each a year, but the thirsty residents of Buenos Aires quaff three times that amount, spending the equivalent of an annual US$140 million on bottled water, most of it home-produced. Two-thirds of Argentina's mineral water sales are in the federal capital and Gran Buenos Aires, which is home to a third of the population. As well as Villa del Sur, the other major Argentinian brands are **Villavicencio** and **Manera**. Most Argentinians prefer their mineral water uncarbonated, and the still varieties outsell the sparkling by four to one. Very recently, however, there has been a significant increase in demand for carbonated water. Perrier and Evian are among the very few imported brands, and they have only a small market share.

VILLA DEL SUR

Villa del Sur: still and highly carbonated, medium mineralization.

Bottling company:
Aguas Minerales SA,
Avenida Juan B Justo 1015,
1425 Buenos Aires,
Argentina.

Analysis:	*mg/l*
Sodium	*154.00*
Potassium	*9.90*
Magnesium	*15.00*
Calcium	*25.00*
Fluoride	*0.91*
Chloride	*34.00*
Sulphate	*23.00*
Bicarbonate	*494.00*

Villa del Sur is Argentina's leading mineral water, supplying at least a third of the total market. In Buenos Aires itself it reckons to have achieved a 50 per cent share. The source is in the plains of Cuenca del Salado, about 100 km (65 miles) south of Buenos Aires. On a wooded estate, three wells tap into the Puelche aquifer, which is filled with water that first fell as rain up to 700 years ago. The water percolates slowly through a complex underground system, passing first into another aquifer, the Pampeano, from which part of the reload volume of water filters downwards to the Puelche aquifer, helped by natural hydraulic charges between the two. Since this process takes centuries, the water matures, picking up trace elements to achieve a unique chemical composition, markedly different from other waters in the area. Thus Villa del Sur is a subtle blend of sodium and bicarbonates, together with magnesium, calcium and a healthy touch of fluoride. The water is extracted from the aquifer at a depth of 70 metres (229 feet), and pumped through stainless steel pipes to the fully automated bottling plant. Just over one-sixth of total production is carbonated.
Production: 120 million litres (31.7 million US gallons) in 1993.

MEXICO

Mexico is really a two-tier market for waters; there are the traditional mineral waters, *agua mineral*, and there is a booming business in purified waters, sold mostly in 5-gallon containers, which grew by 100 per cent just between 1991 and 1993. Over 1,740 companies selling *agua purificada* are licensed, selling more than 80 million 5-gallon containers a year. The biggest purified water group, **Pureza Agua SA**, has bottling plants throughout Mexico; its chief rivals are Electropura in Mexico City, Agua la Purisima in Tijuana and Embotelladora Arco Iris in Guadalajara. These companies, however, are essentially meeting a need for acceptable drinking water.

The mineral waters that you may find in restaurants and hotels across Mexico come primarily from Tehuacán; indeed, **Agua Tehuacán** is almost synonymous with *agua mineral*. The waters of this quiet hill town in Puebla province are sold under several labels, including **Penafiel**, **Garci-Crespo**, **San Lorenzo**, **Etiqueta Azul** and **Balseca**. The hot volcanic springs of Tehuacán were respected by the Indians for their healing powers for centuries before the Spanish conquest of Mexico. They journeyed for many miles to bathe in the highly mineralized waters surging up from the depths of the earth. The Aztec leader Montezuma is said to have taken the water to give him strength to do battle against the invading Spanish army of Cortez.

The water originates from the melting snows on the volcano Pico de Orizabo in the Citlaltepl mountains behind Tehuacán. The water filters slowly down 25 miles to the town's springs, picking up a rich blend of minerals, including calcium, magnesium, sodium and bicarbonates.

The importance of Mexico has also been signalled by the arrival in 1994 of the ubiquitous Nestlé Sources International, which has launched an *agua natural*, **Santa Maria**, in a joint venture with the Mexican group, Manatiales La Asunción. Santa Maria comes from sources in the mountainous region of the famous volcanoes, El Pop Cateptl and El Iztaccihuatl.

National Association:
Asociación de Embotelladores de Agua de la
Región Latino Americana,
Rio de la Plata 2555,
Colomos Providencia CP 44620,
Guadalajara, Jalisco, Mexico.

THE MIDDLE EAST

The desert climates of the Middle East, where temperatures soar to 54°C (130°F) in summer, make countries such as Egypt, Saudi Arabia and the Gulf States natural markets for mineral waters. Copious drinking is essential; daily intake may be at least 6 or 7 litres in the hottest months. Moreover, the sale of alcohol is widely forbidden, so that demand for soft drinks and bottled water is enormous.

The oil rush of the 1970s sent sales soaring. Visiting businessmen and expatriates, working on development projects and often denied a beer, sought consolation with a good refreshing water. Hotel mini-bars were stocked with Évian, Perrier or Vittel from France, San Pellegrino from Italy or Apollinaris from Germany. The boom also triggered a search for local waters.

Actually, Vittel from France was already in the field, going into partnership in 1969 to form Société des Eaux Minérales Libanaises to develop the Ain Sohat source near Beirut (see page 180). Vittel then went on with joint ventures or technical co-operation in Abu Dhabi, Egypt and Kuwait. Indeed, throughout the Middle East it is now quite common to see a line 'with technical collaboration of Vittel' on the label. Évian also followed up, helping with Mineral in Egypt. And local labels have proliferated, usually drawing water from closed aquifers deep beneath the desert. In Saudi Arabia there are at least a dozen bottling companies and in the United Arab Emirates (UAE) there are eight. The UAE is also the largest market in the Gulf, consuming up to 170 million litres (44.7 million US gallons) annually, while the Sultanate of Oman drinks just over 50 million litres (13.2 million US gallons) of local waters. Kuwait, recovering from the war, also consumes over 40 million litres (10.5 million US gallons). The preference everywhere is for still waters.

EGYPT

Egypt is a small market for natural waters, mainly for tourists and foreigners working there, amounting to 30 million litres (7.9 million US gallons) annually. Until the 1980s it was supplied by European mineral waters, but an import tax of 160 per cent was then introduced, making them prohibitively expensive. So the hunt for local sources began, and today's visitor will find four Egyptian labels: **Baraka, Mineral, Helwan,** and **SIWA**.

Baraka is bottled by Vittor, a joint venture with Vittel (now part of Nestlé Sources International), and is the best seller. The water comes from an ancient aquifer located 175 metres (574 feet) beneath the desert north east of Cairo. 'It is a very old water from a closed aquifer that does not renew itself,' Vittel's man in Egypt told us. 'But it will last us at least 100 years. It conforms to all the European norms of purity, constant temperature and consistent mineral content. The three main minerals are sulphate, calcium and magnesium.'

Mineral is Baraka's chief rival, produced with the cooperation of France's Évian. Initially both billed themselves as mineral waters (there being no clear regulations in Egypt), but this definition was challenged in 1984 by a newcomer **Helwan**, a purified water developed with cooperation from Vichy Celestins in France, and Baraka and Mineral now call themselves only natural water (*eau naturelle*). These three waters were joined by a newcomer, **SIWA,** launched in 1993. The challenge for all of them is to lure the Egyptians away from sugary soft drinks, of which they have the highest per capita consumption in the world.

UNITED ARAB EMIRATES (UAE)

The UAE, embracing the great commercial centres of Abu Dhabi and Dubai, is the major consumer of bottled waters on the Gulf. While European mineral waters do have a good niche in the upper echelons of the hotels and restaurants, the volume business is with eight local waters who compete vigorously. They tap aquifers beneath the desert for lightly mineralized refreshing waters to combat the exhausting summer heat. Perhaps the most pleasantly located is at Al Ain, the cool oasis of palms and pools in the hills inland in Abu Dhabi. This is a resort, the home of the Ruler's racing stables, and source of **Al Ain**, natural mineral water from the local springs. Al Ain is bottled with the technical cooperation of Vittel. It is a lightly mineralized water with modest amounts of calcium (7.5 mg/l), magnesium (14.5 mg/l), chloride (22 mg/l) and bicarbonates (75 mg/l).

Rivals include **Emirates** pure spring water with a somewhat higher mineral content 20.2 mg/l, magnesium 30.6 mg/l and sulphates 38 mg/l). This comes from Al Worayaa source in the emirate of Fujairah. Other familier labels include **Masafi,** drawn from an aquifer at a depth of 152 metres (500 feet) in the mountains of the small emirate of Ras-Al-Khaimah. Masafi is a still lightly mineralized with sodium (14.6 mg/l), magnesium (21.6 mg/l) and bicarbonates (62 mg/l).

SAUDI ARABIA

The combination of heat, prohibition of alcohol and the millions of pilgrims visiting the holy cities of Mecca and Medina each year makes Saudi Arabia a natural market for bottled waters, both imported and local. The visitor finds his hotel mini-bar stocked with Évian, Perrier, Sohat (from Lebanon) and assorted local labels. Downstairs in the restaurant almost every table has a big plastic bottle of water in place. The total consumption is 900 million litres (236.7 million US gallons) with a per capita consumption of 60 litres (15.8 US gallons), a remarkably high level until you consider the climate.

Moreover, this is primarily supplied by a well developed local bottling industry; a mere 5 per cent of sales are imported. There are 15 water bottling companies in the Kingdom, all belonging to the Saudi Bottled Water Committee. The chairman, Mahfuz A. Bin Zomah, told us his committee was formed originally in 1989 by the bottlers to consider joint distribution, but instead it has served to cooperate in selecting equipment and exchanging ideas and information. They are planning to extend the committee to include the Arabian Gulf states shortly. The sampling and testing of drinking and mineral waters is controlled by the Saudi Arabian Standards Organisation (SASO).

Among the local waters, the emphasis is very much on providing a pure, healthy water in a desert environment, drawing water from the great aquifers that lie beneath the Arabian peninsula. Indeed, of the waters that we located on visits to Jeddah and Riyadh, no less than nine call themselves 'Health' or 'Healthy' waters.

So if you are in Taif, the cool summer capital in the hills, look out for **Taif Health Water**, and in the commercial city of Damman on the shores of the Gulf, there is **Al-Shifa Health Water**. While **Nissah Health Water** from Wadi Nissah near Jeddah, has wide circulation. The two holy cities of Mecca and Medina naturally have their own waters. In Medina there is the **Madinah Munawarah Health Water Factory**, and from Mecca comes **Makkah**. This appears to be the best seller at 350 million litres (92 million US gallons). **Makkah** water comes from a deep well in a valley in the hills near Mecca. It is lightly mineralized but apparently is treated to reduce mineral content before bottling; it is also ozonated. It is distributed nationally in the major centres of Damman, Jeddah, Mecca and Riyadh.

National Association:
Saudi Bottled Water Committee,
c/o Makkah Water Co,.
PO Box 15164,
Jeddah 21444,
Saudi Arabia.

OMAN

The Sultanate of Oman, at the southern end of the Arabian Gulf, is more mountainous and its wilderness supports much more varied vegetation than its desert neighbours. So the quality of its water is reckoned to be the best in the Gulf, especially inland away from the sea. The National Mineral Water Company, located on the fringe of the mountains in Wadi Kabir, has been able to build a successful business over the last 15 years trading on that reputation. They have tapped into a deep, ancient aquifer with three boreholes through which they are taking off three still waters.

The first is **Tanuf**, launched in 1980, which is relatively highly mineralized compared to other waters of the Gulf or Saudi Arabia, with a good mix of sodium, magnesium, calcium and bicarbonates. A second water, **Jabal Akhdar**, has been bottled since 1988. Although this is from the same aquifer, the mineral content does vary slightly, with a shade more calcium and magnesium, but less sodium. These are the two major labels, but National Mineral Water also markets what it calls a 'popular' water, **A'Ssaha**. The total production for the three waters is 30 million litres (7.9 million US gallons). The waters are all bottled at source in PVC, under standards conforming not only to Omani law, but to European directive 777.

While A'Ssaha is sold only in Oman, Tanuf and Jabal Akhdar are exported in significant quantities to the UAE, Bahrain, Kuwait and Qatar.

Production: 30 million litres (7.9 million US gallons)

Tanuf: still, low mineralization.

Jabal Akhdar: still, low mineralization.

Bottling company:
National Mineral Water Co. (SAOG),
PO Box 2740,
Post Code 112 Ruwi,
Oman.

Tanuf:	*mg/l*
Bicarbonates	210.0
Sulphates	42.0
Chlorides	30.0
Nitrates	8.0
Calcium	52.0
Magnesium	20.0
Sodium	21.0
Potassium	1.6

Jabal Akhdar:	*mg/l*
Bicarbonates	220.0
Sulphates	44.0
Chlorides	28.0
Nitrates	5.0
Calcium	55.0
Magnesium	21.0
Sodium	19.0
Potassium	2.0

Sohat: still, low mineralization.

LEBANON: SOHAT

Travellers to Beirut and, indeed, throughout the Arab world, have long been familiar with the water from Ain **Sohat** - the spring of health - that rises in the Lebanese mountains off the highway between Beirut and Damascus. It is a light, crisp water with a very low mineral content that is always welcome in the searing heat of summers in the Middle East.

The Romans probably knew of this spring near the modern village of Falougha, for their coins have been unearthed by archaeological diggings in the area. Indeed, it was the archaeologist Habib Zoghzoghy who rediscovered Ain Sohat while unearthing Roman remains in the village in 1910. His excavators and the villagers found the water so pure that its fame spread, and bottles of Sohat water, bearing a picture of a Roman coin on the label, were soon being dispatched to the royal courts of Baghdad and Cairo.

Wider distribution through a modern bottling plant, however, came only in 1970 after extensive analysis of the water and its source by Vittel from France. Vittel teamed up with a Lebanese industrial group to form Société des Eaux Minérales Libanaises. This modern Sohat plant taps what is known as the High Spring, slightly above Ain Sohat, on the western tip of Jebel Kneissa at an altitude of 1,710 metres (5,600 feet).

Since the water from rain and melting snow cannot drain off, it percolates slowly down through layers of dolomite and limestone, until it reaches a less permeable bed of marl (limestone and clay) from which the springs are fed through fissures in the side of the mountain. At the High Spring, the water is caught deep inside the mountain in a horizontal concrete gallery in which the spring has been isolated. The flow is a minimum 8,000 litres (2,113 US gallons) per hour at a constant 6.5°C (43.7°F).

During the long years of the civil war in Lebanon, the bottling miraculously continued. On trips to the Middle East, there was Sohat in our hotel mini-bars anywhere from Jeddah to Kuwait. The water is also familiar on the tables of Lebanese restaurants in France and Britain.

Production: more than 100 million litres (26.3 million US gallons).

Bottling company:
Société des Eaux
Minérales Libanaises,
Sohat,
BP 1127006,
Beirut,
Lebanon.

Analysis:	*mg/l*
Calcium	*31.30*
Magnesium	*5.20*
Sodium	*3.50*
Potassium	*0.50*
Iron	*0.01*
Bicarbonates	*105.20*
Nitrates	*1.80*
Sulphates	*10.90*
Chlorides	*5.10*
Fluorides	*0.01*

MOROCCO

SIDI HARAZEM

Sidi Harazem is the best-known Moroccan mineral water, and travellers there will find it offered as the table water to accompany the good local wines served with meals. Morocco, in fact, inherited the tradition of mineral waters from the French during the days of colonial rule, just as it did the pleasant vineyards you see by the road between Rabat, Meknes and Fès. And the custom has remained. Sidi Harazem enhanced its international reputation during the 1980s by winning gold medals at trade fairs in Amsterdam, London and Rome.

The thermal waters at Sidi Harazem – originally known as Kalouan – just outside that dramatic and ancient city, Fès, on the fringe of the the Atlas mountains have been known for over 2,000 years. An oasis of palm-fringed pools amidst the harsh, barren landscape, Kalouan was a desert resting place for centuries. In the 14th century the Sultan Abou Hassan constructed waterfalls and streams around a domed thermal bath. In modern times, at the encouragement of King Hassan II, a new resort complex of hotel and bungalows for those coming to take the cure has been completed.

The water originates from an artesian source 90 metres (300 feet) down through layers of chalk and limestone. The source has now been capped, and the water for bottling flows through a subterranean pipeline to the modern bottling plant built in the late 1960s with technical assistance from France.

Sidi Harazem has a moderately mineralized blend of sodium, calcium and magnesium. It is particularly recommended for kidney and urinary problems. Indeed, brochures state that 'the waters of Sidi Harazem are valued not only for what they bring in [to the body] but for what they take out'. Presumably for that reason, the company warns that the water is not recommended for mixing milk for babies' bottles. Despite that caveat, Sidi Harazem is a very acceptable table water, which is increasingly exported.

Production: not available.

Sidi Harazem: still, medium mineralization.

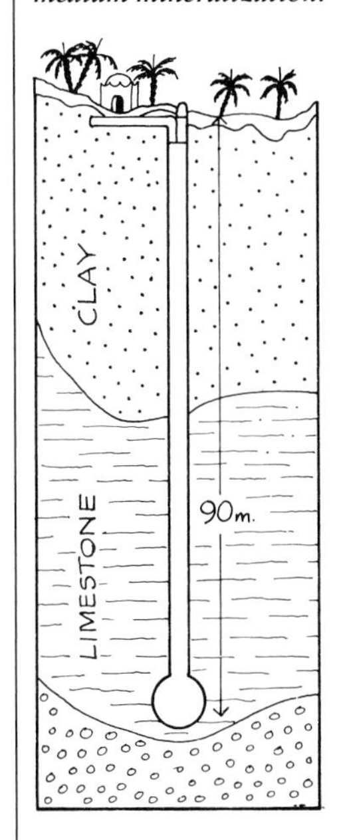

Bottling company:
Société du Thermalisme Marocaine (Sotherma),
BP 210,
Fès, Morocco.

Analysis:	*mg/l*
Calcium	*70.0*
Magnesium	*40.0*
Sodium	*120.0*
Potassium	*8.0*
Chlorides	*220.0*
Nitrates	*4.0*
Sulphates	*20.0*
Bicarbonates	*335.0*

ISRAEL: EDEN

Eden: still and carbonated, light mineralization.

Eden, the most popular brand of mineral water in Israel, comes from the ancient Avel Salukia spring in the north of the country. Far removed from any urban or industrial centre, the spring flows from a deep basalt formation in the Golan Heights, and is one of the sources of the River Jordan. The lightly mineralized water is particularly pure, with a low sodium content, and a palatable blend of calcium, magnesium and bicarbonates.

The spring was given its name in 187 BC by the Greek King Selecus, and its quality was much praised in ancient writings. But after the Roman invasion of the Holy Land during the 1st century BC, the spring was sealed off for 2,000 years until it was relocated in the late 1970s.

The water is now bottled at the source by Eden Springs Ltd in what the company's chairman, Ron Naftali, describes as 'the most advanced plant of its kind in Israel'. The water flows by gravitational force from the basalt rock straight to the automated plant. To its original, non-carbonated Eden mineral water, the company has added a sparkling version, which is available in PET and glass.

The company has also developed a wide-ranging distribution network for selling or leasing electric coolers for the supply of mineral water to offices, stores and other enterprises, and maintains a bi-weekly supply of 5-gallon returnable containers for servicing these units.

Eden's share of the steadily increasing Israeli mineral water market is now more than 60 per cent. The water is guaranteed to be strictly *kosher*, and the bottling plant, which complies with the most stringent international standards, is supervised by advanced internal laboratory and quality control procedures, by outside independent laboratories and by the Israeli Ministry of Health.

Although Eden was initially sold mainly to Israel's tourists and other visitors, local demand for mineral waters has developed considerably over the past 10 years, and the company's annual output has increased 12-fold. Much of its production is still for the home market, but an increasing amount is exported each year, mostly to the United States.

Production: 36 million litres (9.5 million US gallons).

Bottling company:
Eden Springs Ltd,
7 Habarzel Street,
Tel Aviv 69710,
Israel.

Analysis:	*mg/l*
Calcium	*26*
Magnesium	*18*
Chlorides	*24*
Nitrates	*15*
Sodium	*32*
Bicarbonates	*198*

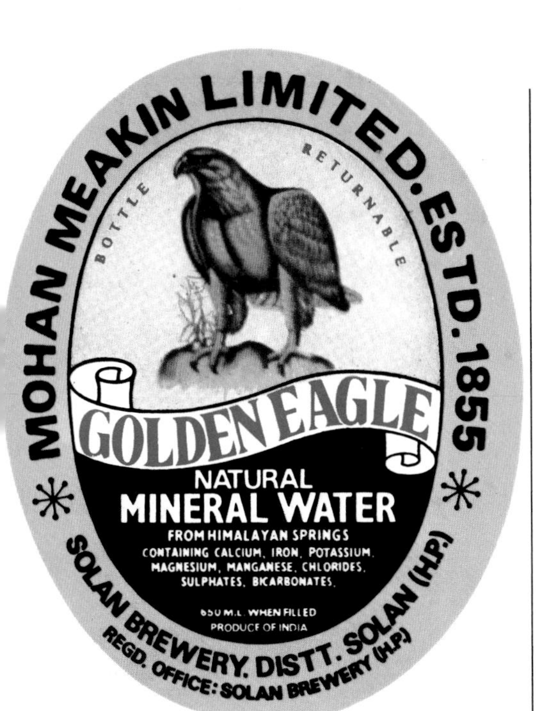

INDIA: GOLDEN EAGLE

Golden Eagle is a pure mineral water from the Himalayas, not especially a local Indian tradition, but a great help to visitors looking for a safe, refreshing drink.

The clear springs on the Karol mountain in the foothills of the Himalayas were discovered 130 years ago, when a local brewery was searching for water. The waters have been used ever since in the production of Solan beers. But in 1977, Brig. Dr Kapil Mohan, managing director of the brewery, decided to bottle the mineral water spring in its natural state to satisfy the growing tourist market. Prof. Wilhelm Schneider of the Fresenius Institute in Germany confirmed the purity and mineral content of the water, and advised on bottling technology. The water and bottling conditions conform to European standards.

The Golden Eagle spring is lightly mineralized, and contains mainly calcium, magnesium and bicarbonates. The low sodium content makes it acceptable to those on salt-free diets. It rises near the watershed between the Indus and Ganges sources, at a height of above 1,520 metres (5,000 feet) in an area of thick sub-tropical vegetation. The waters surface through layers of sandstone, limestone and shales, where they pick up their light mineral content. The modern bottling plant, which was enlarged and upgraded in 1987, can produce nearly 45,000 litres (11,900 US gallons) of mineral water a day, in 500-ml glass and 1-litre PVC and PET bottles for the home and export market. The company plans to produce a sparkling version of the water in 500-ml glass bottles.
Production: approximately 7 million litres (1.8 million US gallons).

Golden Eagle: still, low mineralization.

Bottling company:
Mohan Meakin Ltd,
Solan Brewery P O,
Pin-173214 Shimla
Hills,
India.

Analysis:	*mg/l*
Sodium	1.3
Potassium	2.0
Magnesium	37.0
Calcium	22.0
Chlorides	2.0
Sulphates	34.0
Bicarbonates	196.0

SOUTH EAST ASIA
INTRODUCTION

In the humid heat of South East Asia, whether you are in Bangkok, Singapore, Jakarta or even Vietnam, a glass of cool, refreshing water is essential. Since tap water is often not to be recommended (except in Singapore), bottled waters are essential. Hotels often offer imported Evian, Vittel or Perrier in their mini-bars, but there is an increasing choice of local waters.

THAILAND: POLARIS

Polaris: still.

Bottling company:
North Star Company Ltd,
GPO Box 869,
Bangkok 10501,
Thailand.

The question in Thailand has not been the therapeutic qualities of water, but simply, 'Is it safe to drink?' And the country's best-selling bottled water, Polaris (named after the Pole star), originated entirely from that premise.

Maxine North, the energetic American lady who created the North Star company in the 1950s (the play on her own name is accidental), found that when she first went to live in Bangkok not only she but every newcomer instantly got diarrhoea from the local water, even if it was always boiled. So she conceived the ideal of distributing a purified water from deep artesian wells just outside Bangkok directly to homes and restaurants.

From an initial 15 customers in 1956, Polaris now has 50,000 clients in private houses, hotels, offices, restaurants and supermarkets. You will find Polaris water in all rooms of the distinguished Oriental Hotel and on every flight of Thai Airways International, the national flag carrier.

Besides the two original Bangkok boreholes, Polaris is pumped from similar deep boreholes at three other locations in Thailand, Chiengmai, Hadyai and Korat plants. At all the plants, the water is initially chlorinated, filtered and ozonated to ensure its prime role as a safe drinking water.

Visiting the North Star plant we found a high standard of quality control with many daily laboratory tests on the water at every stage, and all the women workers on the bottling line turned out in crisp white caps and surgical masks. Polaris's success has spawned many imitators in Thailand, but few can offer such attention to hygiene. As one Thai put it, 'Many waters are just from the garden hose.'

Production: 250 million litres (66 million US gallons) in 1993.

MINERE

A newcomer in Thailand since 1993 is Minéré, launched by the Thai entrepreneur, Banyong Pitaksongsanit, in a joint venture, Royal Resources, supervised by Vittel (now part of Nestlé Sources International) from France. Vittel tested sources in Thailand for three years before selecting Pho Sam Ton near Ayutthaya, the former capital of the 18th century Siamese kings. Minéré is a light blend of calcium, magnesium, potassium, sodium and bicarbonates, packaged in PET bottles at a new factory at the source.

Minéré: still, low mineralization.

Analysis:	*mg/l*
Calcium	*54.00*
Magnesium	*27.00*
Sodium	*110.00*
Potassium	*3.90*
Zinc	*0.13*
Bicarbonate	*350.00*
Chloride	*56.00*
Sulphate	*62.00*
Nitrate	*1.30*
Fluoride	*0.12*

SOUTH KOREA: DIAMOND PURE WATER

In South Korea, the main anxiety about water has always been: is it bacteriologically pure and safe to drink? So Diamond Pure Water, the company set up in 1969 never had the intention of bringing a mineral water in its natural state to its customers, but concentrated on treating its spring water in every way to make sure that it was selling a hygienic, sterile product. Climate and history combined to put the emphasis on the need for a bottled water that will do no harm.

Located in the Keimyung Valley, 40 kilometres (25 miles) north of Seoul, the bottling plant stands alongside three wells that have been drilled 130 metres (429 feet) into the local rock to tap a cool 14°C (57°F), clear spring. The water, already pure from its geological filtration, is then subjected to a number of treatments, all in accordance with the World Health Organization's recommendations, and the local Federal Drug Agency's specifications: personnel from the latter organization make random spot checks on the bottling line every two weeks.

The water is collected from the subterranean wells into a tank, where sediment settles and the oxidization process begins. The water then flows through four separate purifying stages, the auto feeder, the clarifier, the service tank and the purifying pump, which remove any possible bacteria. The water is then filtered through a pressure sand filter, an activated carbon filter, and a micro filter.

The end result is a sterile, safe water which is delivered within 24 hours of production to the customer. No stock is ever held. All personnel wear masks for personal hygiene in the plant. The water is delivered in 5-gallon (US) (18.9-litre) polycarbonate containers (used for home or office dispensers), as well as 1.8-litre and 0.9-litre PET bottles.

Production: 36.3 million litres (9.6 million US gallons) annually.

Diamond Pure Water: still, light mineralization.

Bottling company:
Diamond Pure Water Co Ltd,
165-14 Yonhee-Dong,
Suhdaemoon-Ku,
Seoul,
Korea .

Analysis:	*mg/l*
Magnesium	*0.75*
Calcium	*6.80*
Sulphates	*9.00*
Iron	*0.03*
Fluoride	*1.00*
Chlorides	*17.40*

VIETNAM: LA VIE

As Vietnam opens its gates again to foreign investors, one of the first new installations is a modern bottling plant for La Vie mineral water at Khanh Hau source. This is another joint venture supervised by France's Vittel in the province of Tong An on the Mekong Delta 60 kilometres (40 miles) from Ho Chi Minh City. Given France's long association with Vietnam, it is an appropriate alliance.

La Vie is produced and bottled according to European Union standards; it is a light mineral water – not processed or purified like some other Asian waters – with a touch of calcium (23 mg/l). The water was launched in June 1994. The present market in Vietnam is estimated at 30 to 40 million litres (8 to 10.6 million US gallons).

La Vie: still, low mineralization.

Analysis:	*mg/l*
Calcium	*23*
Magnesium	*8*
Sodium	*60*
Potassium	*4*
Bicarbonates	*251*

Our first morning in Sydney, a headline in the *Telegraph Mirror* reported, 'Aussies find it's better by the bottle'. Not, as you might suspect, a bottle of Castlemaine XXXX or Foster's lager, but water. 'We are turning off the taps and taking to the water bottle in search of the fountain of youth,' gushed the story. 'Pollution and an obsession with knowing what we put into our bodies has seen a huge increase in the consumption of bottled water.' The paper quoted Melanie McPherson from the Australian Packaged Water Council saying, 'Water from the tap may be perfectly safe, but it doesn't taste as good. People often buy bottled water simply because they like the taste.'

Actually the bottled water business is just getting started; consumption is still scarcely 10 litres (2.6 US gallons) a head, about where the US was in the early 1980s. And there is often still a delightfully informal frontier atmosphere about it. 'You'll see people by the road north of Sydney just filling up jugs and bottles from a spring,' the marketing manager of one company told us. At many springs where the water is bottled, the operation is small. We called up several in search of information. At one, a chap named Wally answered; he was owner, bottler, marketing manager, and telephone operator; the fount of all wisdom on that spring (and highly suspicious that we wanted to know what he was up to).

But there are good waters, both domestic and imported, around. The imported ones, with sophisticated marketing techniques, are cutting quite an image. At lunch by Sydney Harbour Bridge we drank Crodo Lisiel from Italy and then walked over to the Opera House, where the name of Vittel is prominently displayed as sponsor of productions. On television, in the snappier magazines and on billboards it is the advertisements for Évian that catch the eye. 'We are the

major contributor to advertising in spring waters,' Arno Girard, Évian's man in Sydney, told us, 'and we have over half the market for imported natural springs.' Still water, Girard added, is presently about one-third of the market in Australia, but is growing very fast. Consumption is forecast to double or triple by the year 2000.

Just up the harbourside from the Opera House is the head office of the serious Australian player, Coca Cola Amatil who, besides having the Coca Cola franchise and making soft drinks, owns three of the best-selling local waters, **Deep Spring, Mountain Franklin** and **Taurina Spa.**. They also distribute Perrier and Vittel in Australia and reckon to control up to 60 per cent of the total market. Ian Brown, the corporate affairs manager, neatly summed up the twin track of Australian bottled water for us, 'The Vittel sponsorship of opera is the up-market image,' he said. 'Mount Franklin is what you see at the sports stadium.' 'Mount Franklin,' he added happily, 'is the largest selling still water by light years.'

Mount Franklin was originally named after a spring near that mountain in Victoria, but the water of several other springs is also now bottled under that label. It is a very lightly mineralized natural spring water, with just a touch of sodium 14 mg/l), chloride (20 mg/l), magnesium (5 mg/l) and calcium (3 mg/l). Mount Franklin fits many Australians' perception of bottled water as a pure, refreshing drink after sport. 'The tendency here', explained the marketing manager at Ecks, a Sydney distributor, 'is drinking not so much for minerals as to get a nice taste.'

Another of Coca Cola Amatil's waters, **Deep Spring**, is a sparkling natural mineral water from New South Wales, which has been bottled since 1847. Its mineralization, by Australian standards anyway, is somewhat higher: although sodium (13 mg/l) and magnesium (14 mg/l) are low, calcium is 92 mg/l and bicarbonates are at 168 mg/l. Deep Spring is also available in a range of sparkling flavoured waters.

Taurina Spa is another New South Wales water with similar mineralization. And, like a number of Australian waters, clearly spells out on the label, 'Contains no kilojoules' (i.e. calories). Indeed, as in the United States, the waters are often marketed for

what they do not contain.

The challenger to Coca Cola, especially in Victoria, is Schweppes which, besides overseeing the Évian distribution throughout Australia, bottles a still natural spring water billed as **Diamond of Waters**. This has an exceptionally low mineralization with under 50 mg/l dissolved solids. We sampled it and found that it really did taste as if it was from a spring.

That is the heart of the matter in Australia. Tap water has a poor reputation, especially in Adelaide, South Australia, where a water called **Crystal Springs** is booming in home and office deliveries. The pedigree of a water is less important than a pleasing taste; it is not the historic water of some great spring that is being sold. Indeed, it is hard to find an Australian water that is actually going to bat for a unique spring. The labels often make no mention at all where it is. That is because the waters of one spring may be bottled this year, but another next year. 'Australia is a spring rich country,' a bottler in Sydney said. 'But often they dry up, so it tends to be movable sources.' The hazard of bush fires doesn't help regular supplies either; horrendous blazes just before our arrival had trapped several tankers on their way back from springs to bottling plants, so reserve springs come in handy.

Two other distinctive waters we tried were **Torquay**, a natural spring water from Victoria, which boasts, 'No additives, no calories', and **Bisleri**, a carbonated natural mineral water with a splendid lion's head as a symbol, which begins its pedigree, 'No kilojoules'. We found Bisleri, incidentally, to be one of the pleasantest waters we tasted in Australia

Travelling around Australia, however, you will find a variety of local sources in restaurants and supermarkets. In Tasmania it is **Huon Valley Springs**; in Melbourne look out for **Southern Cross Springs**, **Hepburn Spa,** or **Home Brand Natural Spring**; in Perth you can try **Aussie Natural**, **Black & Gold Spring**, or **Millstream**. They may not appear in the mini-bar of a big hotel. 'The reason,' said a Sydney distributor, 'that you find Évian, Perrier or Vittel in your mini-bar is that the traveller likes to feel comfortable to see international brands (like Coke) and reaches for them. He might be put off by some unknown local label.'

National Association:
Australian Packaged Water Council,
Level 3, William St,
E. Sydney,
New South Wales,
Australia.

NEW ZEALAND

New Zealand is a country of abundant and spectacular water resources – from the hot springs and geysers, boiling mud pools and waterfalls of the volcanic North Island to the rapidly flowing rivers, the calm blue and green mountain lakes, the glaciers and fiords of the Southern Alps which extend the entire length of the South Island. Just the terrain, you might think, for the development of a strong spa culture and mineral water tradition.

But New Zealand is just waking up to bottled waters. The per capita consumption is scarcely 1 litre a year, but the bottlers are looking at what has long been the drinking rate in Europe, and see that even the United States is pressing on towards 40 litres a head.

The New Zealand Apple & Pear Marketing Board, which markets the leading water, **NZ Natural,** is forecasting 'huge growth potential'. But as Dennis White, the market development manager, remarked, the low level of bottled water sales 'is an indication of the high standard of local (tap) water'. So the board is targeting two groups: those who drink for health and, more recently, what they call 'a large group of social alcohol replacement drinkers'.

NZ Natural, which has a 32 per cent market share, is the only major local brand bottled at source. The water comes from an artesian well, at a depth of 45 metres (150 feet). The well is located on the Canterbury Plains of the South Island in the catchment of the Waimahariri river in an area of alluvial silt and gravels comprising hard sandstone. Multiple aquifers occur on these plains, giving rise to artesian wells.

NZ Natural is lightly mineralized with a mere 120mg/l of dissolved solids, including just a dash of sodium and chlorides. It is sold as a still and sparkling water in bottles sizes from 300 ml to 1.5 litres. Its main competitors are **Kiwi Blue** and Évian from France, both of which have around 25 per cent of the market. Thus, between them, the three waters cater for more than 80 per cent of New Zealand's bottled water drinkers.

Bottling Company:
Enza Products,
97 Plunket Avenue,
Wiri, Auckland
P.O. Box 76-202
Manukau City.

JAPAN

Japan is really the final challenge in taking bottled waters world-wide. There is a good way to go. Japanese consumption was only 415 million litres (109 million US gallons) in 1993, a mere 3 litres per head (one-twelfth that of the United States). Yet the demand is growing very fast from a tiny base. It has quadrupled since 1988, and until the recession of 1992/93 was growing by 50 per cent annually. The pace-makers have been imported waters, chiefly from France, but also from the United States, Canada, and Turkey. Their sales have increased faster than local waters, so that imported labels have almost a 17 per cent share. Volvic from France is the favourite imported water, followed by Evian. Both are marketed through the Mitsubishi trading house. Vittel, sold through Marubeni's Sapporo Brewery, and Contrex, Perrier and Valvert, distributed by Nestlé Japan, are also established. But Crystal Geyser from California and Naya from Canada are also making niches.

Among the local waters, we have most frequently encountered **Fuji** (named after Mount Fuji), marketed by Horiuchi, but the overall best seller is **Rokko No Oishii Mizu** – Rokko's Good-Tasting Water – sold by House Foods, which has virtually a quarter of the entire Japanese market. Rokko is a still, lightly mineralized water from the springs of Mount Rokko in western Japan. Suntory, known for its whisky, also has the second best selling water (and is in the bottled water business in the US, too). Kirin and Asahi, famour for beer, also offer waters.

Mineral waters were slow to catch on because of strict legislation to ensure that they were germ-free. Until 1986 all bottled waters had to be sterilized for 30 minutes, quite against the European tradition. It was then agreed that imported waters conforming to European directive 777 need not be sterilized.

Originally, a high proportion of sales were for *mizuwari* (whisky and water) in bars where businessmen keep their own whisky bottle for a quick drink on the way home. "The popularity of bottled water started with *mizuwari*", said Sankichi Yamasaki of Yamasaki International, the industry analyst in Japan, "but now it is not so." Yamasaki reports that more than 80 per cent of sales are now for home drinking, but that is primarily the local waters. The famous imported names retain the lion's share of *mizuwari*, hotel and restaurant sales.

National Association:
The Japan Mineral Water Association,
4 - 17 Kashikawa 2 Chome,
Bunkyo-Ku,
Tokyo 113,
Japan.

Aquamine
Sparkling Mineral Water
EXTRACT OF ANALYSIS FROM CSIR/WATERTEK FROM 28 SEP 93: [mg/ℓ]
CATIONS: CALCIUM (Ca) 57 MAGNESIUM (Mg) 31 SODIUM (Na) 5 POTASSIUM (K) 3
ANIONS: BICARBONATE (HCO_3) 266 SULPHATE (SO_4) 64 SILICA (Si) 11 CHLORIDE (Cℓ) 5
250 mℓ

Aquamine: still and lightly carbonated, medium mineralization.

SOUTH AFRICA

Although the South African mineral water industry is in its infancy, it is a lusty, fast-growing child. Originally, the market consisted of a few home-produced waters, often chemically treated and usually bottled away from source, with some imports, notably from Italy, France and Portugal. But since 1989, a mini-boom has occurred, and virtually all new South African brands entering the market are bottled at source in the European tradition, if not under stringent European-style regulations. Typically, Perrier was the catalyst back in the 1980s with its witty, eye-catching ads, and although lower profile now, it still has a healthy slice of the South African market, where the taste is still largely for carbonated waters. While annual consumption is still only 1 litre a head, demand is growing from young people who associate mineral water with a healthy life style.

One of the pioneers of the South African industry is **Aquamine**, a fresh, moderately mineralized water produced by Barcha Farming Enterprises, 70 kilometres (44 miles) to the north west of Johannesburg. The area is one of rolling dolomitic hills (percolation through which gives the water its moderate mineral mix). Located on Farm Rietfontein in the centre of the Kromdraai Nature Reserve, the Aquamine source and bottling plant are far from possible urban, industrial or agricultural pollutants. You will find Aquamine in South Africa's restaurants and hotels, and in chain stores and off-licences, where still, sparkling and lemon-flavoured versions are available.

Launched in 1988, Aquamine has the distinction of being the first South African mineral water to be bottled at source, and is also the first (and, at present, the only) water whose producers are allowed by the Council for Scientific and Industrial Research (CSIR) to display the results of its official analysis on the label.

Other South African home-produced bottled waters include **Schoonspruit**, which is especially strong in the Transvaal; **Franshhoek Spring Water**, produced by Robin Castell at his mountainside property in the eastern Cape; **Montagu** from Montagu Valley Products in Cape Town; **Caledon**, one of the leaders in the western Cape; and **Absolute Aqua**, marketed by Gilbeys.

Bottling company:
Barcha Farming Enterprises (Pty) Ltd, PO Box 512, Muldersdrift 1747, South Africa.

Analysis:	*mg/l*
Calcium	*57.00*
Magnesium	*31.00*
Sodium	*5.00*
Potassium	*3.00*
Bicarbonate	*266.00*
Sulphate	*64.00*
Silica	*11.00*
Nitrate	*1.20*

CHINA

The Chinese have a traditional respect for the virtues of good water. Indeed, in Cantonese slang the same character stands for both 'money' and 'water' (rather as 'bread' means money in American slang). Early Chinese writings abound with praise of water. The writer Wang Chia of the Chin dynasty noted that, 'The bubbling fountain of Pon Lai gives a thousand lives to those who drink it', while Prime Minister Chang Kun, describing a new spring to the Emperor of the Tang dynasty, observed, 'Chronic ailments of all kinds are cured in no time just by a draught of that water'.

Laoshan mineral waters have a slightly salty after-taste, a little like Vichy water, but they are a refreshing drink in the summer heat. Two waters from the springs of Mount Laoshan on the Shantung Peninsula south-east of Beijing (Peking) are bottled: Laoshan Natural Mineral Water and Laoshan Alkaline Mineral Water. And they are to be found not only in China, but on the shelves of the Chinese Emporium in Hong Kong.

Mount Laoshan itself, now a holiday and health resort, was known in ancient China as the home of the gods, rather like Mount Olympus in Greece. Emperor Shih Huang of the Chin dynasty and Emperor Wu of the Han dynasty are both said to have made pilgrimages to the mountain to drink the water, which had a reputation as an elixir of life. Even today the Chinese point to the excellent health and immunity from disease of their citizens who live in the vicinity of the springs and regularly drink the waters.

Laoshan water was first bottled in 1930. After the revolution a new bottling plant was constructed at Qingdao to bottle the mineral waters and to produce soft drinks. Meanwhile the whole area around the springs has been reforested, and

The Tide's Sound Cascade where all the flows of Laoshan Springs come together.

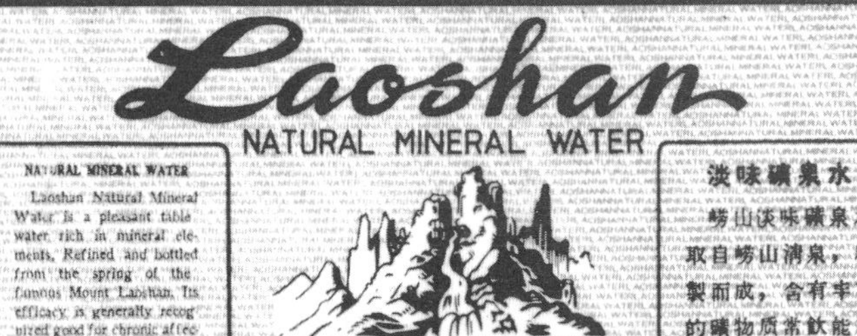

Laoshan: still, high mineralization.

sanatoria for workers to recuperate have been built at Nine North Streams, where many of the brooks flowing from the mountain converge. The surplus waters from the springs themselves tumble into the famous Tide's Sounds Cascade shown in our picture.

The red-labelled Natural Mineral Water is a pleasant table water that is charmingly billed as being 'good for chronic affections of the urinary organs, gastric enteritis, gouty and other troubles of the stomach'. It is also, the bottlers claim, 'nutritive for infants to mix with milk powder or condensed milk'. The stronger Laoshan Alkaline Mineral Water blends calcium and magnesium with surprisingly high amounts of sodium and bicarbonates.

Both waters are exported to Hong Kong (where we drank them), to Singapore and Japan. Several other Chinese waters are also available in Hong Kong, including **Dragon Spring** natural mineral spring water.

Production: Not available.

Bottling company:
China National Cereal Oils and Foodstuffs Corporation,
70 Zhongshan Road,
Qingdao, China.

Laoshan Alkaline:	*mg/l*
Calcium	111.00
Magnesium	70.00
Sodium	1,500.00
Bicarbonates	4,000.00

EUROPEAN UNION MINERAL WATER REGULATIONS

Extracts from directive 777 of the European Union, which was originally issued in 1980. It has now been adopted by most member countries, and is also often taken as a guideline elsewhere.

Definition

"Natural mineral water" means microbiologically wholesome water, originating in an underground water table or deposit and emerging from a spring tapped at one or more natural or bore exits.

At source or after bottling, effervescent natural mineral waters give off carbon dioxide spontaneously and in a clearly visible manner under normal conditions of temperature and pressure. They fall into three categories to which the following descriptions respectively shall apply:

(a) "Naturally carbonated natural mineral water" means water whose content of carbon dioxide from the spring after decanting, if any, and bottling is the same as at source, taking into account where appropriate the reintroduction of a quantity of carbon dioxide from the same water table or deposit equivalent to that released in the course of those operations and subject to the usual technical tolerances.

(b) "Natural mineral water fortified with gas from the spring" means water whose content of carbon dioxide from the water table or deposit after decanting, if any, and bottling is greater than that established at source;

(c) "Carbonated natural mineral water" means water to which has been added carbon dioxide of an origin other than the water table or deposit from which the water comes.

Authorized treatments

1. Natural mineral water, in its state at source, may not be the subject of any treatment or addition other than:

(a) the separation of its unstable elements, such as iron and sulphur compounds, by filtration or decanting, possibly preceded by oxygenation, in so far as this treatment does not alter the composition of the water as regards the essential constituents which give it its properties;

(b) the total or partial elimination of free carbon dioxide by exclusively physical methods;

(c) the introduction or the reintroduction of carbon dioxide under the conditions described in (a), (b) and (c) of the definition above.

2. In particular, any disinfection treatment by whatever means and, subject to paragraph 1(c) above, the addition of bacteriostatic elements or any other treatment likely to change the viable colony count of the natural mineral water shall be prohibited.

Microbiological criteria

After bottling, the total colony count at source may not exceed 100 per millilitre at 20 to 22°C in 72 hours on agar-agar or an agar-gelatine mixture and 20 per millilitre at 37°C in 24 hours on agar-agar. The total colony count shall be measured within the 12 hours following bottling, the water being maintained at 4°C ± 1°C during this 12-hour period.

At source, these values should not normally exceed 20 per millilitre at 20° to 22°C in 72 hours and 5 per millilitre at 37°C in 24 hours respectively, on the understanding that they are to be considered as guide figures and not as maximum permitted concentrations.

At source and during its marketing, a natural mineral water shall be free from:

(a) parasites and pathogenic micro-organisms;

(b) *Escherichia coli* and other coliforms and faecal streptococci in any 250 ml sample examined;

(c) sporulated sulphite-reducing anaerobes in any 50 ml sample examined;

(d) *Pseudomonas aeruginosa* in any 250 ml sample examined.

Without prejudice to the above microbiological criteria and to the conditions for exploitation mentioned below, at the marketing stage the revivable total colony count of a natural mineral water may only be that resulting from the normal increase in the bacteria content which it had at source, and the natural mineral water may not contain any organoleptic defects.

Bottles

Any containers used for packaging natural mineral waters shall be fitted with closures designed to avoid any possibility of adulteration or contamination.

Labelling

1. The sales description of natural mineral waters shall be "natural mineral water" or, in the case of an effervescent natural mineral water as defined above, as appropriate, "naturally carbonated natural mineral water", "natural mineral water fortified with gas from the spring" or "carbonated natural mineral water".
The sales description of natural mineral waters which have undergone the total or partial elimination of free carbon dioxide by exclusively physical methods shall have added to it as appropriate the indication "fully de-carbonated" or "partially de-carbonated".
2. Labels on natural mineral waters shall also give the following mandatory information:
   (a) either the words: "composition in accordance with the results of the officially recognized analysis of (date of analysis)", or a statement of the analytical composition giving its characteristic constituents;
   (b) the place where the spring is exploited and the name of the spring.

Commercial designation

1. The name of a locality, hamlet or place may occur in the wording of a trade description provided that it refers to a natural mineral water the spring of which is exploited at the place indicated by that description and provided that it is not misleading as regards the place of exploitation of the spring.
2. It shall be forbidden to market natural mineral water from one and the same spring under more than one trade description.
3. When the labels or inscriptions on the containers in which the natural mineral waters are offered for sale include a trade description different from the name of the spring or the place of its exploitation, this place or the name of the spring shall be indicated in letters at least one and a half times the height and width of the largest of the letters used for that trade description.
4. Paragraph 1 (above) shall apply with regard to the trade description used in all other forms of advertising relating to natural mineral waters.

Restrictions on labelling and advertising

1. It shall be forbidden, both on packaging or labels and in all forms of advertising, to use designations, proprietary names, trade marks, brand names, illustrations or other signs, whether emblematic or not, which:
   (a) in the case of a natural mineral water, suggest a characteristic which the water does not possess, in particular as regards its origin, the date of the authorization to exploit it, the results of analyses or any similar references to guarantees of authenticity;
   (b) in the case of drinking water packaged in containers which does not satisfy the provisions relating to a natural mineral water, are liable to cause confusion with with a natural mineral water, in particular the description "mineral water".
2. (a) All indications attributing to a natural mineral water properties relating to the prevention, treatment or cure of a human illness shall be prohibited.
   (b) However, the indications listed in the table below shall be authorized if they meet the relevant criteria.

Indications and criteria relating to the composition of the water, including references to special diets

Indications	*Criteria*
Low mineral content	Mineral salt content, calculated as a fixed residue not greater than 500 mg/l.
Very low mineral content	Mineral salt content, calculated as a fixed residue, not greater than 50 mg/l.
Rich in mineral salts	Mineral salt content, calculated as a fixed residue, greater than 1,500 mg/l.
Contains bicarbonate	Bicarbonate content greater than 600 mg/l.
Contains sulphate	Sulphate content greater than 200 mg/l.
Contains chloride	Chloride content greater than 200 mg/l.
Contains calciumx	Calcium content greater than 150 mg/l.
Contains magnesium	Magnesium content greater than 50 mg/l.
Contains fluoride	Fluoride content greater than 1 mg/l.
Contains iron	Bivalent iron content greater than 1 mg/l.
Acidic	Free carbon dioxide content greater than 250 mg/l.
Contains sodium	Sodium content greater than 200 mg/l.
Suitable for a low sodium diet	Sodium content less than 20 mg/l.

BIBLIOGRAPHY

Acque Minerali Italiane 93-94, Edizioni Laus, sas, Milan, 1994.
L. M. Crismer, *The Original Spa Waters of Belgium* (SA Spa Monopole NV, Spa, 1983).
Emmanuelle Evina, *Le Guide du Buveur d'Eau* (Solar, Paris, 1992).
Timothy Green, 'Apostles of Purity', *Smithsonian*, 15, 7 (October 1984), Washington DC.
W. S. Holden (ed.), *Water Treatment and Examination* (J. & A. Churchill, London 1970).
Le Marche des Eaux Minérales Naturelles en Europe 1983-1992 (GISEM-UNESEM, Paris, 1993).
Milan Mlacnik (trans.), *Three Hearts At Your Fingertips* (Radenska and Mladkinkska Knjiga, Ljubljana, 1982).
S. M. Bel'enkii, G. P. Lavreshkina, Tat'iana Nikolaevna Dul'neva, *Mineral Waters* (Moscow, 1982).
Steven Schwarta, *The Book of Waters* (A. & W. Visual Library, New York, 1979).
Douglas A. Simmons, *Schweppes, The First 200 Years* (Springwood Books, London, 1983).
Alberto Violati, *San Gemini e Carsulae* (Bestetti, Milan, 1976).
Arthur von Wiesenberger, *Oasis* (Capra Press, Santa Barbara, 1978).

ACKNOWLEDGEMENTS

Cover: Perrier UK. **Title Page:** Chris Warde-Jones. **5** Ralph Crane. **7** Reproduced by permission of *Punch*. **9** Ralph Crane. **11** Timothy Green. **13** *The Independent*/Herbie Knott. **14** Timothy Green. **21-23** Timothy Green. **25** Ralph Crane (2). **27-35** Timothy Green. **39** Ralph Crane. **41** courtesy Sangemini. **43** Chris Warde-Jones. **44-45** Timothy Green. **47-49** Chris Warde-Jones. **51** Ralph Crane. **67-69** courtesy Apollinaris. **73** courtesy Fachingen. **75** courtesy Gerolsteiner. **76-77** courtesy Hassia & Luisen. **79** courtesy Überkinger. **85** courtesy Spa Monopole. **86-87** Timothy Green. **93** courtesy EBAMSA. **98** courtesy Melgaço. **100-101** Timothy Green. **105** courtesy Valser. **107** Timothy Green. **109** courtesy Valser. **112-115** Timothy Green. **119** Antony Jones/UK Press. **141** courtesy Ramlosa. **151** courtesy São Lourenço. **156-157** Timothy Green. **192-193** courtesy China National Cereals, Oils and Foodstuffs.

INDEX TO WATERS

E

F

G

H

I

J

K